WITHDRAWN

The Weiser Field Guide to
cryptozoology

Werewolves, Dragons, Skyfish, Lizard Men, and
Other Fascinating Creatures Real and Mysterious

D1089074

ⓦ WEISER BOOKS
San Francisco, CA / Newburyport, MA

First published in 2010 by
Red Wheel/Weiser, LLC
With offices at:
500 Third Street, Suite 230
San Francisco, CA 94107
www.redwheelweiser.com

Library of Congress Cataloging-in-Publication Data
Budd, Deena West.
The Weiser field guide to cryptozoology : werewolves,
dragons, skyfish, lizard men, and other fascinating creatures
real and mysterious / Deena West Budd.
p. cm.
ISBN 978-1-57863-450-7 (alk. paper)
1. Cryptozoology. I. Title. II. Title: Field guide to cryptozoology.
QL88.3.B83 2010
001.944—dc22
2009049352

Front Cover: moon, night monster, sea creature, snow monster, dragon
gargoyle © istockphoto.com. Wolf eye, creature in tunnel, and gargoyle face
© dreamstime.com. Illustrations © BB Faye and James Budd.
Illustrations on pages 12, 18, 34, 43, 51, 63, 67, 76, 90, 93, 120, 126, 135, 138, 140,
143, and 150 © BB Faye, and James Budd. Images on pages 17, 20, 37, 44, 49, 60,
65, 74, 75, 76, 86, 97, 100, 102, 105, 108, 123, 124, 130, 134, 141, 145, 146, 159 x
3 and 166 © dreamstime.com. Images on pages 6, 9, 13, 19, 22, 28, 30, 46, 53, 54,
57, 59, 71, 81, 83, 89, 95, 111, 116, 136, 154, and 161 © Dover Publications.
Images on pages 25, 29, 33, 119, and 164 © istockphoto.com.

Production Editor: Michele Kimble
Proofreader: Nancy Reinhardt
Typeset in Jenson and Priori

Printed in Canada
TCP
10 9 8 7 6 5 4 3 2 1

*Dedicated to my own faithful
little four-legged golden-haired cryptid,
Goku Buu Budd*

Contents

Introduction

While writing this book, I was astounded to discover that many people had never even heard the word *cryptozoology*. I was also amazed to find that many dictionaries and online word origin sources did not include the term. I did find a definition at *AskOxford.com:* "**noun** the search for animals whose existence is disputed or unsubstantiated, such as the Loch Ness monster." Most cryptozoologists say it is "the study of hidden animals." The term is credited to the "master of cryptozoology," Bernard Heuvelmans and dates back to 1959.

Cryptids, the name for the creatures studied in this field, are not only beings whose existence has not yet been proven; the name is also given to creatures sighted in areas to which they are not indigenous. Cryptozoology can include the study of beings that have been considered extinct for thousands of years.

There is debate among researchers and cryptozoologists regarding what creatures can be called cryptids.

Many of the experts investigate only hairy humanoids, but some fringe cryptozoologists accept mystical creatures such as unicorns, fairies, and dragons. Traditional beasts such as vampires, werewolves, and zombies are often included in the spectrum as well. Ghostly entities, such as black dogs and shadow people, are sometimes found under the cryptid heading. We can't leave out the often-sighted aliens, which are believed to be the source of a great number of cryptid sightings.

It was not an easy task deciding what cryptids to include in this book. I started out researching the more well-known creatures such as Big Foot, the Loch Ness monster, and the hairless beast mentioned often in the news lately, the *Chupacabra*.

Oh my goodness, the information available on Big Foot is amazing! Hundreds of thousands of volumes and online sources to read and research. In addition to Big Foot sightings recorded in most states, countries, and continents, there are reports of similar creatures using many different names. Some cryptozoologists classify other cryptids, such as the Ohio Grassman, the Beast of Bray Road, Goatman, and the New Jersey Devil, as Big Foot-type creatures.

The gray areas in the field of cryptozoology are immense at this time, as it is a relatively new area of study. There are thousands of hidden creatures that have been sighted all over our awesome world. Were you aware of the giant kangaroo sightings in the Midwest in the 1970s? Have you heard of the giant turtle living in

Indiana? Did you know about the little boy in a small town in Illinois who, while playing hide-and-seek in his backyard, was picked up and carried 40 feet by a giant bird?

In addition to these documented creatures, there are many incredible cryptids that we don't normally hear about. I decided I might serve my readers best by delving into these lesser-known creatures, like the shocking Mongolian Death Worm, the Awful, the goblins of Kentucky, the Loogaroos vampires, the Ahool, and many more incredible and marvelous beasts. I've also included some of the cryptids that have been in the news over the years, such as the Boston Lemur, the Chupacabra, Mothman, and the giant rats of New York City.

One of the most fascinating types of creature sighted in contemporary times is the rod or skyfish. I enjoyed researching this cryptid immensely, and am grateful to expert Jose Escamilla for his time and help on the subject. Thousands of articles, books, and videos have been written and produced by many talented, brave, and persistent researchers and cryptozoologists. Because of their often dangerous treks into the unknown, their travels to the locations of reported sightings, and their diligence in recording the information, I was able to write this field guide to help the novice cryptid hunter.

Look upon this book as an introduction to forty different incredible beasts. I have tried to provide the information you need to know: where the creatures come from, where to find them, what traits and characteristics to look for, and how to prepare yourself for the search. At the end of the book is a Further Reading section, in which I list many of the sources on each cryptid. I have included references I used in my own research and also the sources of my sources!

I've looked at not only many cryptids from the United States, but also fascinating creatures from all over the world. Some can be found in your yard or even inside your own home! Many of these creatures are benign and would be delightful to encounter. There are others, though, that I've delineated more as a warning than as an enticement. I implore you to use great caution while searching for these hidden animals. Keep in mind the cautions I include for the more dangerous cryptids. The most important bit of advice I can give to you is to remember not to show your fear; most evil creatures gain their strength from our weaknesses. Appearing brave will give you courage. Good luck!

Rods (Skyfish)

It's difficult to imagine that there might be creatures flitting about in the air around us, ordinarily undetectable by the naked eye. Yet many people believe that organisms, often called skyfish or rods, flying snakes, or serpents, and occasionally referred to as solar entities or atmospheric beasts, are swimming or undulating through the air around us right now.

These amazing cryptids have been captured on film numerous times in various places around the world, even under water. They were first discovered in this medium in 1994 by producer/director Jose Escamilla while shooting a documentary in Midway, New Mexico.

In November 1996, at a deep pit called Solano de las Golondrinas, in the Cave of Swallows in Mexico, some rods were inadvertently videotaped by Mark Lichtle, who was recording some BASE jumpers. It was not until after several trips to the area, when a segment of the Lichtle video aired on television, that Escamilla noted the rods on the video. A television crew from San Diego TV station KFMB accompanied Escamilla to the Cave of Swallows in 1999 and filmed what might be a colony of skyfish.

Escamilla believes the anomalies captured are living creatures that dart through the air at speeds so fast that they are barely discernible by the untrained human eye. Escamilla (actually, his wife) coined the term *rod* because of the creatures' cylindrical shape.

Some observers contend that the rods are three-dimensional and seem to operate in an intelligent or instinctual manner, controlling their own flight paths. Often appearing playful, chasing each other around and even interacting with birds. Rods sometimes come out of nowhere for a fraction of a second, only to disappear.

Rods range in size from a few inches to as long as 100 feet. Some have appendages that resemble fins or wings along the length of the cylinder. Many times, multiple sets of wings are observed, or a thin membrane of wings is wrapped around the length of the body. It is reported that the torso undulates as it travels. Escamilla identifies the three types of rods as "centipede rods," which have several pairs of fins or wings; "white rods," which appear to have no fins and are more ribbon-like in shape; and the colorful "spears," which can be yellow, white, or brown. These last ones are thinner and faster than the other types of rods and do not have fins.

Theories about the origins of rods vary considerably; their sources are thought to range from aliens and interdimensional beings to atmospheric beasts. Some think they are some kind of secret military weapon. There is one theory that they might be distant relatives of the anomalocarids or "strange shrimp," early marine animals believed to be extinct.

Although Jose Escamilla is credited with being the "discoverer" of these flying creatures, other reports precede Escamilla's; there are 1,000-year-old carvings in Argentina that resemble rods or skyfish.

Although Escamilla believes that with patience and practice we can learn to see rods with the naked eye, most are detected on video. Some scientists speculate that our ancestors might have had better eyesight than we do because of our modern habits of reading and watching television. Our ancestors' eyes would have been better trained for self-protection and to see far distances.

Accounts from China dating back to A.D. 747 describe serpents flying in the air. In September 1891, in the town of Crawfordsville, Indiana, a flying "serpent" was seen by several witnesses. It was "swimming" about 100 feet in the air. It appeared to be 20 to 30 feet in length and 8 feet wide, with at least one pair of wings or fins attached.

Trevor James Constable, author and researcher, photographed what he called "critters" or "sky creatures" in the 1950s using infrared film. Skyfish have been filmed by photographers for the Discovery Channel, the History Channel, and the Learning Channel.

During the filming of the movie *Braveheart* in Scotland, a rod was seen in the blue sky passing by Mel Gibson's head. They've also been captured in a music video made in Mexico.

Escamilla has collected more than 2,000 images of rods from as far back as a 1910 sporting event. He has footage, taken in 1957 by a naval official with a 35mm camera, that shows the cryptids leaping straight up out of the ocean into the sky. Escamilla was amazed at their speed, estimating that they travel from 150 to possibly 1,000 miles per hour! Escamilla often observes the skyfish high in the air, streaking by at "extremely high velocities."

On an overnight flight from Dallas to Denver, CO, Escamilla claims he saw a dark rod entering the cabin through the airplane window. He saw it pass in front of people as it "darted down the aisle toward the front of the plane."

In May 1999, a rod was observed flying through a rare and violent F5 tornado near Oklahoma City. It seemed to emerge from a cloud. Or perhaps it was "phasing" in and out, appearing and disappearing, materializing and dematerializing. This is the sort of behavior that might be typical of an interdimensional entity. However, most rod sightings are of them flying, not popping in and out of space.

In late October 2002, Brandon Mowry, a photo-journalist for Fox News in Albany, New York, saw an odd cylindrical object zooming past an airplane while he was editing some tape he had taken at the airport. He was very curious about what he was seeing. Was it a missile, or an unidentified flying object (UFO)? Mowry notified airport security, but it had not been caught on radar. The Federal Bureau of Investigation (FBI) was contacted. An agent interrogated Mowry and confiscated the tape. No further comment or reply has been heard from the FBI regarding this evidence.

In 2003, a rod was seen in Baghdad after a huge explosion of a Swedish tank at the test firing range, leading to speculation of a possible connection between rods and secret military weapons. Many rods have been spotted near military bases, operations, and aircraft.

On a July 31, 2003, episode of *Coast to Coast AM* with George Noory, Escamilla made the claim that there is new evidence of skyfish from the United Kingdom. He tells of a home video showing a rod flying into the open mouth of a girl singing at a wedding. He claims she is seen to pull it out of her mouth! Escamilla also tells of footage of how a football referee is shown to flinch when a rod runs into his body at a game.

At a Minnesota Zoo in 2005, an ape was recorded who seemed to be aware of a rod nearby, although the photographer did not notice it.

If you know what to look for, and you have enough patience, it is possible to observe skyfish without the use of any video equipment. When our pets react to something that we aren't seeing, Escamilla says it could be a rod. He believes that when we see something out of the corner of our eye, it is often a rod rather than an insect or bird. Speculation that the creature might be a bird or an insect has led to the examination of the Escamilla video by experts in those fields. Ornithologists have determined that there are no birds resembling the entity found on the video.

Professor Wooten from the University of Exeter in the United Kingdom states that the way the rod propels itself through the air does not fit in with any known method. He believes that it is "not inconceivable" that a rod might be some kind of unknown bug, but there is "nothing particularly insect-like about the images."

One very odd finding is a rod with three pairs of wings discovered by Escamilla. We are not aware of any insect in existence with three pairs of wings.

Suggestions regarding the lack of evidence of rod carcasses bring up the possibilities that their bodies are absorbed up into the atmosphere, that they instantly decay upon death, or that they morph into something unrecognizable.

Escamilla offers some tips on how to catch the creatures with video equipment: It is easier and less costly to capture the skyfish on video cameras rather than still cameras. Adjust the camera's shutter to the higher sports setting. When you aim your camera at the sky, be sure to include a frame of reference, like a tree. Don't use a wide-angle zoom. Adjust it so that it looks like what you are seeing with your naked eye. Be patient. Escamilla recommends recording for at least ten minutes before readjusting angles. When you play back your video, if you see an anomaly, slow the speed down to replay.

There is a company in Japan that advertises "Spoodles," traps for catching rods. Amazon Japan sells a DVD entitled *How to Catch the Skyfish*.

For me, the jury's still out on this one. There just isn't enough information available at this time to make any determination as to what these anomalies might be, but if they do exist, and they are "alive," my question is: What are they eating?

Mongolian Death Worms

Although they're not nearly as large as the worms in the movie *Tremors,* a rooftop would still be as good a place as any to come across the acid-spitting, electricity-throwing Mongolian Death Worm. Of course, a rooftop might be rather difficult to find in the Gobi Desert.

Approximately 1,000 miles by 500 miles, the Gobi Desert is one of the largest deserts in the world and has more rocky areas than sand and dunes. Considered a "cold" desert, its temperatures range from freezing at night to sweltering in the daytime.

An inhabitant of the southern Gobi Desert, the Mongolian Death Worm is first mentioned in 1926 by Professor Roy Chapman Andrews, a paleontologist, in his book, *On the Trail of Ancient Man*. Mongolia's nomadic tribesmen called the creatures *Allghoi* or *Olgoi-Khorkhoi*, which means "intestine worm" or "blood-filled intestine," because that is what the cryptid resembles.

The Mongolian Death Worm is usually described as blood red with darker spots, although it has been said to change color to match its environment. About 2 feet in circumference, it ranges from 2 to 5 feet in length. The invertebrate doesn't appear to have eyes, a nose, or a mouth, so it is difficult to tell its head from its tail. Spiked projections appear at both ends.

Czechoslovakian author/explorer Ivan Mackerle believes the creature hibernates most of the year, only coming out of its underground chamber in the hot, rainy months of June and July. The worm moves in a sideways motion across the desert floor, similar to the sidewinding motion of some desert snakes.

The worm is considered to be extremely danger-ous because of the toxic acid-like substance it spits as a defense mechanism. It is easier to tell which end is its head when it is spitting at you! It also has the ability to discharge a lethal jolt of electricity from several feet away. For this reason, the worm has been compared to an electric eel. But electric eels don't spit and they don't live on land. Some spectators state that the worm "raises half of its body up" and "inflates itself," emitting a "bubble of poison from one end." The poison corrodes everything it touches, even metal. However, the toxin loses its effectiveness as the creature's time of hibernation approaches.

Often seen in the vicinity of the saxaul plant, it has been considered that the worm might obtain its poison from the plant's noxious roots or from the goya plant parasite found on the roots.

A witness currently working as an interpreter for an exploration team remembered an incident from his childhood when a visiting geologist was killed instantly by a "huge fat worm" that emerged from the ground. A native ranger tells a story from the 1960s of an entire herd of camels killed by a worm lying below the surface of the desert. The locals tell the story of a worm hiding inside a yellow toy box and killing a little boy instantly when he reached inside. It then killed the child's parents when they tried to exact revenge.

There is another tale about two friends riding on horseback on a hot July day. One fellow and his horse both suddenly fell down dead. The other fellow saw a "big fat worm slowly crawling away."

To the west of Mongolia, in the neighboring country of Kazakhstan, the Death Worm is called *bujenzhylan*. Similar worms have been reported in other countries, although they aren't known to be dangerous.

The *Megascolides australis* of Victoria reaches lengths of 13 feet; the *Didymogaster sylvaticus* of New South Wales is a "squirter earthworm" that spews harmless internal fluids into the air out of its pores; and a *Microchaetus rappi* of South Africa found in 1936 measured 22 feet!

Michel Raynal, a cryptozoologist from France, suggests that the worm might be some kind of burrowing serpent or cobra, although the creature is said to have smooth blotchy skin, not scales.

 Similar to an earthworm, the cryptid makes its way to the Earth's surface after a rainfall. It also appears to respond to terrestrial vibrations. Usually, though, an earthworm requires a damp, moist climate, not the arid environment of the desert.

During the past few decades, a few exploration teams have ventured into the Gobi searching for the Mongolian Death Worm. They have used various techniques, such as bucket traps, ground "thumping," and sending shock waves through the ground in attempts to roust the creature out into the open. A team led by Dr. Chris Clark in 2005 was driven out by a sandstorm. Prior to that, following a rainstorm, Dr. Clark had found that "the whole desert floor was covered in burrows."

This is one cryptid I'd advise not trying to find. The Gobi Desert is full of ticks, biting flies, and vicious spiders; that's enough to deter me!

Ahool
and Other Giant Bats

I have always loved bats. I have one tattoo, and it is a little bat on my shoulder. Maybe my infatuation with these flying mammals has something to do with my lifelong love of vampire movies and exploring caves. Of course, I might not feel so affectionate toward them if I encountered a bat with a 12-foot wingspan and a face like a monkey!

The Indonesian island of Java contains numerous volcanoes and some amazing cave systems. Usually, where there are caves, there are bats. A giant bat called the *Ahool*, named for the sound it makes, has been sighted numerous times throughout western Java since 1925, when it was first seen by the naturalist Dr. Ernest Bartels.

Using the claws on its featherless wings, the Ahool is able to capture large river fish for its food. It is said to be dark gray in color, with a flat face that looks like a monkey's (a macaque or a gibbon) and huge black eyes. The form of the Ahool's feet indicates that the creature is likely to hang upside down, as most bats are known to do.

Cryptozoologist Ivan T. Sanderson believes the Ahool is related to the species of insect-eating bats called *Microchiroptera*. Further, he thinks the Ahool is an Asian version of the Zambian *Kongamato* or the *Olitiau* of Cameroon.

Although there are similarities, the Kongamato is not quite as big as the Ahool; it has reddish fur, and its snout is long rather than flat. *Kongamato* means "breaker or overwhelmer of boats," and the flying pterosaur-like creature is said to attack small boats and is considered to be extremely dangerous, according to Frank Melland in his 1923 book, *In Witchbound Africa*.

In 1956, in what is now Zambia, an engineer spotted two Kongamatos flying quietly through the sky. The creatures circled around and flew overhead again, allowing Mr. J. P. F. Brown a thorough look at them. In addition to the standard description of the Kongamato, Brown noted a long, thin tail, narrow head, and a "mouth full of sharp teeth." In 1957, near the same location, a man with a bad chest wound showed up at a hospital, saying he had been attacked by a creature fitting the description of the Kongamato.

The *Olitiau* ("forked one" or demon) of Cameroon looks much like the Kongamato, although its body fur is black and its wings are blood red. It has large serrated white teeth, a 12-foot wingspan, and a monkey face. Cryptozoologist Ivan Sanderson encountered the cryptid bat near a mountain stream in 1932 when it dived at him before flying off.

Madagascar, an island near Africa's coast, is said to harbor a giant bat called the *Fangalabolo* (very enjoyable to say), which has a wingspan of more than 5 feet. This bat likes to glide down from the sky and tear the hair from people's heads (not enjoyable to encounter).

One of the most feared giant bats is the *Guiafairo* ("the fear that flies by night") of Senegal in West Africa. This "smelly" cryptid hides in hollow trees and caves during the day, but it has been known to invade people's homes at night. It is gray in color with clawed feet.

Less frightening is the *Mlularuka* of Nietoperz, Tanzania, which has been around for a very long time. The size of a small dog, it is a fruit eater and a pest to farmers.

According to the African Ashanti mythology, there is a horrid batlike creature from southern Ghana, Togo, and the Ivory Coast called the *Sasabonsam*. It is reported to be the size of a man, hairy, with a wingspan of nearly 20 feet. The Sasabonsam has defined ridges above its eyes, and long teeth. The creature is also said to have an emaciated body and twisted legs.

Vampirish in nature, the Sasabonsam sits in trees waiting for its victim to pass below it. The beast then pounces on its victim and sucks his or her blood. It has been rumored that a Sasabonsam was killed and photographed in 1928, but if so, no evidence remains.

The *orang-bati* ("men with wings") are said to be from Indonesia—specifically, the island of Seram. These creatures are approximately 5 feet tall and rather feminine in appearance. They have bodies the color of blood and black fur on their wings and long tails. Sounding very much like the winged monkeys of *The Wizard of Oz*, they fly through the air at night raiding small villages, snatching babies and small children, before returning to their home in an extinct volcano to eat their catch.

A giant bat called the *Batsquatch* has been sighted several times since 1980 and is purported to live on Mt. Saint Helens, in the state of Washington in the United States. With skin the color of an eggplant, eyes the color of blood, huge bat wings, and a loud, deep yell, the Batsquatch would be an amazing creature to encounter. They are believed to feed on livestock, as large quantities of animals in the area disappear frequently.

A huge, flying cryptid from Papua New Guinea, called the *Ropen* or *Indava* (also known as a pterosaur or pterodactyl), glows as it glides through the nighttime sky. Mainly subsisting on fish, it is said at times to have resorted to grave robbing and eating human flesh.

An expedition to seek out the Ropen in Papua in 2006 recorded on video what might be two Ropens. One was also observed in the daylight, sleeping under an overhanging cliff. Even more recently, Joshua Gates of the SciFi series, *Destination Truth*, has recorded luminescent images in the sky thought to be Ropens.

In 2007, Jonathan Whitcomb, author of *Searching for Ropens*, interviewed a woman from South Carolina who saw a creature that fits the description of a Ropen flying over a highway near a swamp. Ms. Wooten told Whitcomb that the beast "looked as big as any car, and had no feathers" with a wingspan of approximately 15 feet.

Surprisingly, the giant bat is one of the cryptids sighted most frequently, with reports coming in from all over the world. In my exploration of caves, I have had several encounters with bats. Once while crawling through a cave hole at dusk (that was not a good time to be in that location—what was I thinking!), I heard a tremendous roar and had  to cover my head as hundreds (maybe thousands) made their way out through the small tunnel to feed. One landed on my shoulder for a minute, which did give me a bit of a fright. A few years ago while I was walking past a restaurant in Quincy, Illinois, a bat flew out the open door and right up my skirt! The restaurant owner was mortified, but I am proud to say I was quite calm. I even took the little fellow to some limestone caves in the area to be with the rest of his colony. Yet I doubt I would remain calm if a bat the size of an Ahool flew up my skirt!

Loogaroo
Vampires of the West Indies

My favorite horror tales as a child always included
the romantic stories of the long-suffering vampire.
Hurrying home each afternoon after school to watch
the television show *Dark Shadows*, with the devilishly
handsome Barnabas Collins and the beautiful but evil
Angelique, I dreamed of being a vampire. And there are
all sorts of vampires. . . .

In the Caribbean islands, there are said to be vampires who were once old humans, usually women called
"hags" or witches, who had made deals with the Devil.
According to natives, they are called *Loogaroo*. Satan
gives the hags certain magical powers in exchange for
providing him with blood. Other island names for the
bloodsuckers include *Asema, Sukuyan, Nigawu,* and
Aziman.

The witches are said to gather at the silk cotton or
Devil's tree each night, remove their human skins, and
transform into balls of fire, sometimes blue in color.

In the West Indies, when people see a flash of fire,
they are certain it is a Loogaroo. When natives awaken
without any enthusiasm or energy, they have no doubt
that a Loogaroo has been at work over them during the
night, draining their life's blood and their vital essence.
Sometimes, the victim's blood is particularly tasty, and
the Loogaroo will be unable to stop until the victim is
drained dry.

Similar to the traditional belief of a vampire perishing if he is unable to return to his coffin at dawn, the Loogaroo suffer terribly and die if they can't retrieve their skin to regain human shape before daylight. If you should happen to come upon a Loogaroo skin, sprinkle it generously with salt and pepper. The hag will then be unable to use the skin without developing terrible sores that make her easily identifiable as a Loogaroo. If she declines to put the skin on, she will not become human again and will perish. Stretching the skin so that it doesn't fit the hag properly also works.

Animals are particularly vulnerable to the attack of a Loogaroo. Pet and farm animal owners are encouraged to take extra precautions. Dogs can be infected by the "vampire taint," which changes your pooch into a pet you won't want to be around. Horses are especially vulnerable to attacks. All holes and windows of stables and animal pens should be covered with metal netting to deter Loogaroo. Garlic spread around the area is also said to work against these West Indies vampires. Unlike the traditional theory that vampires must be invited into one's home, doors and windows don't stop the Loogaroo.

The magic extended to the Loogaroo comes at another cost to the creature. Oddly compulsive in habits, the Loogaroo has an instinctive need to "count." Therefore, throwing grains of rice, seeds, peas, nails, or other items on the ground will stop the Loogaroo from pursuing you, as it will feel an overwhelming desire to stop and count the items.

For many years, I have known about voodoo and have heard stories of zombies in the West Indies being controlled by a *bokor*, but I had no idea that vampires are so prevalent in the beautiful islands. Sometimes beauty does exact a price.

Bunyips (Water Horses)

Being a naiad myself, I find water cryptids to be fascinating. One water creature with some especially interesting theories on its existence is called the *Bunyip*.

In Australian aboriginal folklore, the Bunyip is considered to be aggressive and dangerous, with supernatural characteristics. According to the legends, the Bunyip was a water spirit who lived in all the waters of the continent. Anyone who ignored the frightening thunderous booming sound of the monster and approached the home of the Bunyip was most likely eaten.

The Bunyip that was sighted throughout the 1800s and its rarely sighted contemporary relation appear to be of a less dangerous variety than that of aboriginal legend, preferring to eat grasses and herbs.

Although the Bunyip ("bogey," "devil," or "spirit") is first documented in the 1820s, a reference in 1812 to a *Bahnyip* ("seal-like") creature may have been the initial modern citation. Cryptozoologist Bernard Heuvelmans theorized that the origin of the word *Bunyip* is the aboriginal *buynil* or "Supreme Being."

Sounding similar to the Celtic Kelpie, a water horse that inhabits the lochs of Ireland and Scotland, this cryptid might be related to the Orkney *nuggle*, the Swedish *Backahasten*, the Norwegian *Nokken*, and the Scottish *Each Uisge*.

The amphibious monster is about the size of a small calf or pony. It resembles a sheepdog, with black (although some accounts say white) shaggy hair, but it has been described with wings, fins, flippers, claws, walrus tusks, alligator scales, and feathers. The variety sighted most often has a bulldog-like face. Another variety is

called "long-neck," having a more extensive neck and a horselike mane and tail. This variety has been reported only in the area of New South Wales. Often seen swimming, it has been reported to run in an "awkward, shambling gallop" when necessary.

Throughout the 1800s, the Bunyip was sighted regularly in lakes and rivers in New South Wales, Victoria, and Australian Capital Territory, as well as in Tasmania. In late autumn of 1821, a dog-faced Bunyip with dark black hair was sighted in a marsh near Lake Bathurst, New South Wales, by E. S. Hall. In 1846, on the banks of Murrumbidgee River in New South Wales, a strange skull was found. Experts in the vicinity concluded that it was "the skull of something unknown to science." A year later, the skull was put on display at the Australian Museum in Sydney for two days. An overwhelming number of visitors flocked to the museum to see the skull and, according to the Sydney *Morning Herald*, many witnesses were encouraged to speak out about their own Bunyip sightings. Not long after being put on display, the skull disappeared from the museum.

In 1847, a cow herder looking for missing stock in a flooded area saw a long-necked variety of the Bunyip grazing in a pasture. According to a report made to the Sydney *Morning Herald*, the creature was the size of a large calf, dark brown in color, with two large walrus-like tusks, a long neck, and a long pointed head. The Bunyip's ears were very large and seemed to prick up when it realized that it had been spotted. A thick mane of hair ran from the head down the neck. It had large forequarters in proportion to its hindquarters and a thick tail. The creature appeared to be frightened by the observer and awkwardly shambled away. Next morning, broad and square tracks were observed in the muddy ground.

A dog-faced variety of the Bunyip was spotted in 1852 in Lake Tiberias, Tasmania. It was approximately 4½ feet tall, with a bulldog head and black shaggy fur. In another sighting, Charles Headlam was rowing across Great Lake in Tasmania with a friend when they nearly collided with a dog-faced Bunyip. It was the size of a fully grown sheepdog, with two small flippers. It stayed on the surface of the water until it had swum out of view.

A description given to a newspaper in 1872 by one of three men watching a dog-faced Bunyip in Midgeon Lagoon, New South Wales, for half an hour described it as "half as long again as an ordinary retriever dog" with long, jet black, shiny hair all over its head and body. No eyes or tail were discerned, but the ears were "well marked."

Horsemen fording a river in 1886 near Canberra reported a dog-faced Bunyip the size of a dog, with a white coat. It disappeared after the men threw rocks at the creature. At about the same time, a similarly described creature was shot at in New South Wales. It made an odd grunting noise and withdrew into a lagoon.

In 1890, the Melbourne Zoo sent an expedition to capture a Bunyip seen often in the Euroa district near Victoria. The expedition failed.

In 1932, sightings of Bunyips were reported from the hydroelectric dams in Tasmania. Recent sightings have been reported from the area of Lake George and Lake Bathurst.

Are these creatures descended from the similarly described prehistoric diprotodons (giant sloth), which are said to have been extinct for 10,000 years? Or, perhaps are they descended from some other type of megafauna that has been extinct for thousands and thousands of years? Are they seals that have been misidentified? Or, as some speculate, are the more current sightings merely glimpses of fugitives or "swaggies" (outlaws) that have been hiding in the swamps and billabongs?

Kentucky Goblins

I've known some people who have reminded me of goblins—one boss I can think of, in particular. I'm fairly certain, though, that this was just my warped perception of these individuals.

But one never knows! In western Kentucky, in the Kelly and Hopkinsville region, a family was once attacked by a group of goblin-like creatures.

It was a hot, late evening on August 21, 1955. At their rural farmhouse, the Sutton family members were visiting with their guests from Pennsylvania, the Taylors. The sky was just beginning to darken when Billy Ray Taylor walked outside for a cool drink from the well. He noticed something odd in the sky shooting out flames of all colors. It then appeared to stop and quickly descend into a ravine a few hundred yards behind the house.

Billy ran back inside to report what he had seen, but no one else took it seriously enough to investigate. About an hour later, the Sutton dog began to bark excitedly. Billy Ray Taylor and Elmer Sutton walked outside to see why their canine was causing such a ruckus. About this time, the dog crawled under the house, where he remained the rest of the night.

Taylor and Sutton saw a glowing figure coming toward them from the trees. It appeared to be a male, about 3 or 4 feet tall, with metallic-looking skin. It was humanoid and very slender with thin appendages. It had claws for hands, a large, bald head with batlike ears, and a long slit for a mouth. The entire creature was glowing, including its eyes. The skinny arms were raised in the air: the creature's legs appeared to have wasted into uselessness, and it moved in a floating manner, using its arms as ballasts.

As the beast came nearer to the porch, the men became frightened and grabbed their rifles. They later agreed that they had shot the creature numerous times. It just somersaulted backward and floated off into the woods. The boys then shot another creature on the porch roof as it was reaching for Taylor's hair. As it fell, it seemed to float to the ground. Another goblin was at the side window. Another in a tree. For hours the humanoids kept appearing and the men kept shooting.

One of the adults at the farmhouse, Mrs. Lankford, asked that the men stop shooting, as the creatures hadn't hurt anyone and the children were hysterical. The men disagreed, feeling that the creatures seemed to be purposely attempting to scare the bejesus out of them! Finally, worried for the children, the families loaded them into their cars and got out of there. They headed to Hopkinsville for help and arrived close to midnight.

Upon their arrival at the police station, it was obvious to Police Chief Greenwell that they were terrified of something. Corroborating their story was a report from a state trooper near Kelly of "unusual meteors flying overhead" during the time their battle with the goblins was taking place.

Police officers, military police from Fort Campbell, reporters, and photographers descended on the farm. They noted the bullet holes and used shells. They saw a green light in the woods with no apparent source. They even saw an unexplainable greenish tinge on the fence where one of the goblins had been shot. Unfortunately, a sample was not taken.

The investigators and media eventually left, with inconclusive results. The family again settled in, trying to get some sleep. Immediately the goblins were back, peeking in at them through the windows! Finally, around dawn, the creatures disappeared.

Investigators and media involved in the incident came away from it with no real evidence but also with very little doubt that the witnesses were telling the truth about what they'd seen. Their stories were consistent. Eventually tired of being hounded by tourists and other thrill seekers, the family now avoids talking about the goblins. In 2002, however, Elmer Sutton's daughter, Geraldine Hawkins, stated that she had no doubt her father was telling the truth.

According to archives of the Mutual UFO Network, extraterrestrial sightings were recorded in the United States that night and the following in the areas of Chalmette, Louisiana; Casa Blanca, California; Roseville, Michigan; Woodlawn, Ohio; and, North Platte, Nebraska.

In 1998, Karal Ayn Barnett stated in her article, "The Kelly-Hopkinsville Incident—A Historical Review," that from her experience in the region, no one would gain anything at all from making up such a story and would be subject to "ridicule," "contempt," and hatred from the community. Ms. Barnett further indicated that is the reason the families moved from the area soon after the goblin attack.

The story of the goblins of Kentucky brings to mind an incident of my own that occurred in 1985. I was living alone in an apartment in Little Rock, Arkansas. One night, I awakened to the sound of something large scurrying through my home. I immediately thought of the Zuni fetish warrior in the movie *Trilogy of Terror*. I had no animals and no rodent problem (and it sounded much bigger than a rodent). I lived in a secure apartment complex. I could not imagine what this thing could be running around in my living room. I didn't want to find out, so I waited. Eventually, I heard the creature come into my bedroom. I kept my eyes closed but could feel it looking at me for a minute or two. Then, it ran out of the room and seemed to leave my apartment. When I got up the next morning, there was no sign that anything had been scampering around inside.

Tommyknockers of California

As a lover of caving, or "spelunking" as many like to say, the Tommyknockers could be very helpful to me.

I have had the pleasure of exploring a few old abandoned mines. As a kid, I was lucky enough to have a wonderful granny who lived close to some abandoned limestone mines. She was awesome about letting all of us kids explore as much as we liked. Those caves provided the foundation for many unforgettable adventures.

When I lived in the western region of the United States, I enjoyed exploring old ghost towns built during the era of the great gold rushes. Many of these abandoned towns in the middle of nowhere include old mines to explore as well.

Originating in the Cornwall region of England, the Bucca Boo fairies were brought to the United States by the Cornish in the 1820s, when they immigrated into western Pennsylvania to work the coal mines. In the United States, they became known as Tommyknockers. When the California Gold Rush took off, the miners and their Tommyknockers made their way westward.

Said to be approximately 2 feet tall and green in color, Tommyknockers are often compared to leprechauns and brownies. In Germany, they are called *Kobold, Berggeister,* or *Bergmannlein,* meaning "mountain ghosts" or "little miners." They are known by many names in many cultures, including the Manx *Buggan,* the Irish *Pooka,* and the Welsh Bwci. In these other cultures, in which fishing and farming are more dominant than mining, the fairies are just as helpful in these occupations.

Sometimes the little mining fairies like to play small jokes and be ornery, as fairies are wont to do, but usually their purpose for hanging around was to inform the miners of trouble. They would knock on the walls of the mine to warn the workers of an approaching cave-in.

Many miners believed that the Tommyknockers favored them with gifts and good luck. But, there are always a few bad apples in any bunch, and these fairies brought disaster and death to some of the miners. There were mines that had to close because of the evildoings of the Tommyknockers. The little men were said to stay in the area of the closed mines, moving into the homes near the mine shafts, and wreaking havoc on the families living there.

Usually the miners looked favorably upon the Tommyknockers; though they would blame the creatures for missing hammers, they would just as readily thank them for their help. In addition to warning of collapses in the mines, the Tommyknockers also helped with other mining duties, working along with the men. Oftentimes, the miners would leave bits of food and little gifts for the creatures.

The Cornish believe the Tommyknockers contain the souls of the Jews who crucified Christ and were sent by the Romans to work as slaves in the tin mines. As time went on, this belief seemed to change somewhat, and Tommyknockers were often looked upon as dead miners' souls.

Cornish immigrants were insistent about not entering the mine to work until they were assured by the mining company that Tommyknockers were in residence. In 1956, the closing of a huge mine in California prompted descendants of the original Cornish immigrants to petition the owners of the mine to "set the knockers free, enabling them to move on to other mines." The owners met the petition's demand.

Enfield Horror (Giant Kangaroos)

The children's television series *Captain Kangaroo* was a favorite of kids all over America from 1955 to 1984. I loved watching it through the '60s. In 1973, a kangaroo-type creature not nearly as friendly as the Captain, made its appearance in the Midwest.

On a late April night in 1973, Mr. and Mrs. Henry McDaniel of Enfield, Illinois, returned home to find their children, Henry and Lil, excited and frightened because something had been scratching at the door, trying to get into the house.

Later that night, the beast returned, and Henry was ready for it, armed with a pistol. When Henry opened the door, he couldn't believe what he was seeing. He described the creature as having a grayish color, about 5 feet tall with a short trunk, three legs, two short little arms extending out of its chest, and two huge pink eyes!

McDaniel shot at the creature several times, certain he had hit it at least once, but it bounded away, hissing "like a wildcat," in leaps that cleared 50 to 75 feet each. It disappeared along some railroad tracks.

State troopers called to the McDaniel place found tracks "like those of a dog, except they had six toe pads." The investigators also learned that a child playing in his yard behind the McDaniel home had been mildly attacked by the creature, who stepped on the boy's feet, destroying his tennis shoes.

As the McDaniels talked about their experience with others, curiosity seekers descended on the small community, angering the sheriff. He jailed several of the monster hunters who reported that they had also seen and fired upon the beast.

A few days later, Henry McDaniel was awakened by howling dogs in the very early morning hours and saw the creature near the railroad tracks close to his house. The beast was walking calmly along the tracks.

After hearing the story, news director Rick Rainbow from WWKI radio station in Kokomo, Indiana, visited Enfield with some friends. At one point, they saw the animal by an abandoned house near the McDaniel home. It was running away from them, but they could see it was about 5 feet tall, although hunched over, and gray in color. Mr. Rainbow did get a tape recording of its cry.

Researcher, author, and expert cryptozoologist Loren Coleman heard the recording of the high-pitched wailing sound. He had heard the same cry himself when he investigated the creature near the McDaniel place in Enfield. (*See* Further Reading for more information.)

For months after that, screams were heard in the woods and creek bottoms. When the creature killed a dog, the townspeople tried to track it, but the ability of the beast to "leap from 20 to 40 feet in a single bound" enabled its rapid escape.

The kangaroo/monkey-like creature seems to like to travel around quite a bit, especially through the Midwest. After appearing in Enfield in 1973, it headed for Chicago. A marsupial attacked two Chicago police officers when cornered in an alley in October 1974. It got away by leaping over a high fence. The creature was sighted several more times in the Chicago area over the next couple of months.

In the summer of 1899, a kangaroo was seen in New Richmond, Wisconsin. In 1934, farmers in South Pittsburg, Tennessee, complained about a giant kangaroo killing dogs and fowl. It was tracked to a mountainside cave but then disappeared. Coleman found a record of a sighting in the summer of 1941 of a similar-sounding beast in Mt. Vernon, Illinois. A man was out squirrel hunting when a large baboon-type animal jumped out of a tree near him. The hunter fired his shotgun and the creature left.

In 1949, a bus driver in Grove City, Ohio, saw a creature meeting the description of the kangaroo-type beast. In Nebraska during 1958, many sightings were reported. All through the late 1950s and the 1960s, sightings occurred in the woods near Coon Rapids, Minnesota. The creature was reported in Abilene and Wakefield, Kansas, in 1965; Puyallup, Washington, in 1967; Route 63 in Ohio and at Michigan State University in 1968; back in Kansas in 1971; Oklahoma in 1975; Golden, Colorado, in 1976; Wisconsin again in 1978; up into Canada in 1979; and, then down to Delaware.

In 1980, the only sighting reported was in the San Francisco area, but by the following year, reports came in all summer in Utah, Oklahoma, and North Carolina. The creature was seen near Detroit in 1984 and in Iowa in 1999.

As a Missouri girl, I've been looking, but haven't found any reported sightings of Giant Kangaroos in the Show-Me State. How curious . . . maybe they're afraid of MoMo! (*See* **A Tale of MoMo.**)

Frogmen of Loveland

One of my favorite science fiction B movies is *The Creature from the Black Lagoon* from 1954. That's how I picture the Frogmen of Loveland. I discovered during my research that the first sighting of these cryptids occurred in 1955, the year after the movie was released. Nevertheless, the story bears telling.

A businessman traveling through the little town of Loveland, Ohio, at 3:30 a.m. one morning in May 1955 witnessed three humanoid/reptilian creatures by the road. He pulled over and watched for a few minutes, until one of the beasts started waving something that looked like a Fourth of July sparkler around in the air. The businessman decided at that point that it was high time to continue his road trip.

He did provide a description: approximately 3½ feet tall with "leathery skin" and "webbed hands and feet." Their heads were distinctively "froglike," with deep grooves instead of hair. The eyes were bulging, and the mouths were wide.

Five months later, in August, a woman swimming in the Ohio River near Evansville, Indiana, was attacked by something below the water. It grabbed her and tried to pull her under. She got away, but was covered with long scratches and bruises. There was also an extraordinary greenish handprint on her leg.

Seventeen years later, very early on the morning of March 3, 1972, a policeman traveling down an icy road saw a creature dash in front of his vehicle, nearly causing him to crash. After the policeman stopped his vehicle, he saw a frogman in his headlights, squatting down on two legs, looking at him. After a quick moment, the frogman jumped over the guardrail and down to the Ohio River.

The policeman provided description similar to the businessman's: 3 to 4 feet tall, weighing between 15 and 75 pounds, with skin that looked leathery and "features resembling those of a frog or lizard." When another officer checked out the scene later, he did find "scratch marks" on the guardrail.

Two weeks later, another officer, driving toward Loveland, spotted a frogman on the road. At first,

 thinking the creature was dead, the officer got out of his vehicle to remove the roadkill from the highway. The frogman jumped up, scaring the officer into drawing his revolver and taking a shot. The hybrid, not taking his eyes off the public servant, escaped via the guardrail and river once again. The description of the frogman in this account matched the others, with the addition of a tail.

Investigators uncovered records of many more sightings in the area by farmers and other local residents. From eyewitness accounts, it is known that the frogmen are excellent swimmers and agile leapers.

In 1985, two boys in the area claimed to have seen a dog-sized frog by the river. Other sightings of similar creatures include an incident in Pretare d'Arquata, Italy, in 1993. To see some fascinating pictures and read a young man's captivating and informative account, please see *ufocasebook.com*. This witness had several encounters with the beings. He touched one of the monsters with his foot, and the foot turned black for several days. A large number of headless and limbless hens were found in the area during this time, although curiously, there were no reports of any blood around the poultry.

There was another sighting in Varginha, Brazil, in 1996, witnessed by more than eighty people; two of these creatures were captured. One of the policemen who captured a creature with his bare hands died less than a month after the incident due to a bacterial infection. There is speculation that it might have been unintentionally transferred from the being. (*See hyper.net* for an interesting video.) An unusually high number of animals in the zoo in the area died mysteriously during this time.

In an interview in 2003, a Brazilian doctor claimed to have examined one of the creatures and noted reptilian skin, "three bony prominences" on the top of the head, a very long and forked tongue, as well as external veins. He also asserts that the feet ended in a type of claw.

The Miami, an American Indian tribe, has in its folklore tales of an amphibious creature in the Loveland area centuries ago. Many investigators believe that this creature is similar to the *kappa*, or river imp, of Japan, and the *Chupacabra* of Latin America.

Many times, while floating on my back, gazing up at the sky from the surface of a lake, quarry, river, or other body of water, my wicked imagination starts to spin some heinous scenario: something smelly, slick,

and vicious rises up beneath me, snatches me with
its sharp claws, and pulls me down to the icy depths
below. I'm sure the scene will now be even more clearly
defined in my head. Yikes!

Gray Man of Ben MacDhui

I love adventure. I love that adrenaline rush from bun-
gee jumping, caving, white-water rafting, and skydiving.
I was the first radio personality in my area to skydive
while live on air. However, one adventure that I have
not yet had the pleasure of experiencing is mountain
climbing. If I ever do find the opportunity, I don't think
I'll start with the Highlands of Scotland!

In the crests and passes of Ben
MacDhui, the highest peak of the
Cairngorm mountain range (well over
4,000 feet), in the eastern Highlands
of Scotland, lives a 10- to 20-foot-tall
creature. Known as *Am Fear Liath
Mor* in Gaelic, or simply "the Big
Gray Man," this creature is difficult
to describe.

Sometimes he appears to be very
tall, broad, and hairy, a wild man or
Big Foot–type of being that walks
erect. He is frequently said to be a
presence more often felt than seen.

For hundreds of years, there have been sightings of similar creatures called *Wudewas* or "Wood Men" in other areas of the United Kingdom. Well into the sixteenth century, images of these creatures were depicted by engravers in coats-of-arms. The images are also included in many medieval paintings.

According to the hundreds of documented sightings of these wild men, they are usually observed carrying wooden clubs, tree branches, or crude bows and arrows.

In 1925, an explorer and scientist, John Norman Collie, spoke at a meeting of the Cairngorm Club in Aberdeen. He told the members of the club about his experience while descending Ben MacDhui thirty-four years previously. Something huge was stalking him down the mountain. He could hear the creature's footsteps crunching through the snow behind him. He became very frightened and ran to the Rothiemurchus Forest several miles away. He refused to go back to the mountain again.

Other mountaineers came forward with stories about similar frightening experiences on Ben MacDhui. Although Professor Collie did not see the entity, many others described a gray, huge, hairy humanoid with an abnormally large head and neck. There were also reports of music, singing or humming, laughter, inexplicable voices, and footsteps on the mountain. One sighting included clawed feet and pointed ears in the description. Another claimed the beast was wearing a top hat.

What seemed to dismay the mountain climbers the most was the feelings of fear and panic that accompanied the sightings. The feelings were nearly overwhelming and seemed to entice them into becoming less careful and confident, even suicidal, as they attempted to descend the high peak. After reaching a particular place in the descent down the mountain, the negative feelings were said to disappear quickly.

Most witnesses believe the Gray Man has some kind of supernatural power enabling it to instill negative feelings in the observer. Some accounts report that the creature is in control of the fog and mountain mists.

The *Brenin Llwyd* ("Gray King") also known as the Monarch of the Mist, lives in the mountains of Snowdonia in Wales and is very similar to Am Fear Liath Mor. King of his mountain, he lives alone, waiting for gullible travelers and lost hikers.

One interesting theory is that the mountains inhabited by these beasts are portals to other dimensions. Perhaps these Gray Kings are from another dimension and act as guards to the doorways.

Nearly twenty-five years ago, I talked with a fellow traveling through Little Rock, Arkansas, who claimed to be on his way to a portal or doorway to another dimension. He told me he knew that he would have to fight the guardian of the doorway and that it wouldn't be easy. He said that he planned to write a book about his experience, if he made it back. As a voracious reader, I spend much time in bookstores. I have not yet found this young man's (middle-aged by now, I guess) story. I wonder if he ever made it back!

Lusca and Other Giant Octopuses

Since the release of the classic movie *Jaws*, beachgoers have been leery of swimming in the deep-blue sea. I have to agree that the jaws on that shark were awfully big and full of teeth! I'm certain, though, that I would be just as terrified if a long, pink tentacle wrapped itself around my leg and started pulling me down into the depths of the ocean.

An amazing creature, the octopus can camouflage itself by changing its skin color. Its hues can also shift according to its mood! Comprised of mostly muscle, and having no bones, the beast is able to squeeze into surprisingly small places. Red octopuses of Puget Sound, Washington, enjoy making their home in beer bottles. Half of the creature's brain neurons are in its multisuckered tentacles. If separated from its body, a tentacle can crawl on its own. Octopuses' diet consists of just about anything, even each other!

One of the first documented sightings of a giant octopus occurred around Thanksgiving 1896. Two bicycling boys discovered an enormous light pink mass that had washed up on the beach at St. Augustine, Florida. Although only partly visible in the sand, what could be seen was measured as 18 feet long and 7 feet wide. When Dr. DeWitt Webb examined the creature, he was convinced that the mass was a giant octopus. Recent research has led biologists to back up Dr. Webb's identification. Photographs were taken of the octopus; unfortunately, they no longer exist, although there are sketches of the carcass.

A local resident found several octopus-like "arms" near the body. The longest one was said to measure 32 feet.

Correspondence between Dr. Webb and a scientist colleague, Dr. A. E. Verrill (discoverer of the giant squid), indicates that Dr. Verrill eventually determined the specimen to be a giant squid rather than an octopus. In the January 1897 issue of the *American Journal of Science* and an 1897 issue of the magazine *Nautilus*, the creature was referred to as a squid. However, the January 3, 1897, issue of the *New York Herald* indicates Verrill had stated the creature was a portion of an octopus that weighed 20 tons, with a tentacle span of 200 feet. An article in the February issue of the *American Journal of Science* corroborated this verdict, naming it *Octopus giganteus.* Squid or octopus, it would be something to see as you're bicycling along on a sunny afternoon by the beach. I doubt that it made much difference to the boys who first discovered the body.

A few weeks after its initial discovery, the monster returned to the sea from which it came. Then it washed back up again on another beach a short time later!

Using four horses and six men, Dr. Webb tried to turn the giant octopus over but was not very successful. He was able to move the beast higher up the beach, discovering that it was actually 21 feet long. The body was then opened up, and the internal organs removed. The organs were smaller than expected and "did not look as if the animal had been so long dead." The tissue was very hard and extremely difficult to cut.

Dr. Webb encouraged Dr. Verrill to travel to Florida to examine the body of the beast, but he was unable to do so. Dr. Webb did send some samples to him, which Verrill received February 23. In the article published in the *New York Herald* on February 14, 1897, Dr. Verrill hypothesized that the tentacles would have been more than 100 feet long. He went on to speculate that the creature may have been injured or killed by a sperm whale, partially eaten, and then washed up on shore.

There has been no documentation found to indicate what ultimately happened to the giant octopus, and it wasn't until 1957 that Forrest G. Wood, searching through some old files of the Marineland Research Lab in Marineland, Florida, came across the newspaper clippings. The only remains left of the creature were some tissue samples at the Smithsonian. These samples were examined by scientists at this time, who verified the giant octopus theory.

Fishermen in the Caribbean have many tales to tell of giant scuttles (Bahamian for "octopuses"). The commissioner of Andros Island tells of fishing with his father as a small boy when a giant octopus attached itself to the bottom of their boat.

The giant octopus or scuttle is reported to have a 200-foot "tentacle span." It prefers to live in deep water, sea caves, and possibly in the blue holes of Bimini, and makes its way into shallow water only if it is sick. These octopuses are only dangerous to fishermen if the creatures can manage to hang on to a boat with one tentacle while holding on to the ocean floor with another.

While working in the Bahamas, Mr. Wood talked to an expert fishing guide about scuttles. The guide told him about several sightings. The most recent one had occurred in 1946.

Over the years there have been other discoveries of the carcasses of "globsters," which many scientists believe are the same beasts as the St. Augustine giant octopus. An octopus measuring 20 feet by 18 feet was found in western Tasmania in 1960. In 1965, a "globster" 30 feet long and 8 feet high was found on Muriwai Beach in New Zealand. Another globster was discovered in 1970 on a beach near Sandy Cape, Australia. It was 8 feet long. In 1988, in Mangrove Bay of Bermuda, another 8-foot-long carcass was found on the beach. The carcasses eventually washed back out in the water.

The Lusca of Andros Island is said to live in the blue holes of Bimini, hundreds of feet deep, with the largest a quarter of a mile in diameter. The clear waters of the holes reveal stalagmites and stalactites, indicating that they are an immense network of underground caves formed during the Ice Ages. These small freshwater lakes connect to the Atlantic Ocean.

The Lusca has been described as incredibly large and unbelievably fast, having the jaws of a shark and the tentacles of an octopus. Usually referred to as a female, she is often blamed for lost ships. Some locals state that the Lusca steals people off of ships as well. Divers into the blue holes in 1984 reported seeing giant shrimp and crabs there, which would meet the dietary needs of a giant octopus. An attempt to trap these creatures and bring them to the surface was met with resistance from below! Something very heavy kept breaking the traplines when the divers would try to raise them. At one point, the divers' boat appeared to be "dragging," and they saw "a pyramid shape approximately 50 feet high" on his sonar. A few minutes later, something began to drag his boat "along at a speed of around one knot." When the diver put his hand close to the trap rope, he felt "thumps like something was walking, and the vibrations were traveling up the rope." Suddenly, the rope loosened, the trap was raised, and it was found to be bent on one side. The divers have no doubt that the culprit was a giant octopus.

Local witnesses have also described boats of fishermen being pulled beneath the water in the blue holes. Pieces of boat float to the top, but there is never a sign of the sailors.

Kraken, often compared to the Leviathans from Job in the Holy Bible, are giant tentacled sea monsters off the coasts of Norway and Iceland. Erik Pontoppidan, Bishop of Bergen in the mid-1700s, in his book, *Natural History of Norway*, describes this beast as "the size of a floating island." When diving back down into the waters of the ocean, it left an extraordinary whirlpool that often sucked the ship down after it!

Often, though, the creature preferred to grab on to the craft, even large warships, with its magnificent arms and pull it down. The waste of the Kraken is said to feed armies of small fish, tempting fishermen to take the risk and fish in the vicinity of the monster.

Many speculate that the giant squid in the waters of Norway and Iceland, known to grow to 43 feet in length, might be the Kraken of legend. They live in the deepest recesses of the water but have reportedly attacked ships at times.

Pierre Denys de Montfort, scientist and researcher in the study of mollusks, claimed in his book, *Histoire Naturelle Générale et Particulière des Mollusques*, in 1802, that there are two types of octopus in existence: the Kraken and another larger type that attacked sailing vessels off the coast of Angola from Saint-Malo.

In the 700-foot-deep waters of Puget Sound, south of the Tacoma Narrows in Washington, lives the Giant Pacific Octopus, normally ranging from 5 to 6 feet in length, and having a 10- to 12-foot tentacle span. Local legends claim a 200-foot octopus frightened sailors for hundreds of years in this region.

In 2005, an underwater photographer was attacked by a 50-foot octopus that took his camera away from him. The largest identified Giant Pacific Octopus was 600 pounds, with a tentacle span of 30 feet.

Scott Cassell, world-class diver and underwater cameraman, was part of the *MonsterQuest* Investigation Team that included R. J. Myers and Dale Pearson.

Cassell had previously had an altercation with a giant octopus that had a beak similar to a parrot's, about 4 feet in diameter.

While he was growing up, Myers heard a legend about an octopus measuring 40 to 60 feet living beneath the ruins of the original Narrows Bridge. During one of Myers's dives, he found evidence of an octopus den in an underwater cave. In 2002, the U.S. Department of Homeland Security shut down the divers.

With 95 percent of the world's oceans still unexplored, I have no doubt that many more amazing cryptids will be discovered in our briny seas.

Flying Man of Falls City, Nebraska, and Other Flying Humanoids

I once read a short story in an English class about a winged manlike being who, I believe, had injured a wing and landed in a farmyard chicken coop. He was then made to endure living in the coop while the landowners charged a fee for tourists to come look at the unfortunate being. I thought it was a very odd story and didn't much like it. I had no idea at that time that such creatures might really exist!

On an autumn afternoon in 1956 in the town of Falls City, Nebraska, a "man" with wings was seen flying about 15 feet above the ground. The wing apparatus was attached to the man via a shoulder harness. It was

made of a gleaming metal mate-
rial with lights of many colors
lining the underside of the
sheeny wings. They spanned
15 feet and were controlled by
dials on a panel attached to the
man's chest. The man appeared
to be wizened and old, with large
runny eyes and a devilish face. The witness was
immobilized as the creature flew over his head.

A story in the New York *Sun* from 1877 tells of the
sighting of a winged humanoid over Brooklyn, New
York, in September. In 1880, several witnesses reported
a flying man with an angry expression on his face above
Coney Island.

Similar reports were announced in Voltana, Spain,
in 1905 by more than two hundred witnesses. This
time the humanoid was a female dressed in white. She
did not have wings; she was flying against the wind,
and might have been singing. In Portugal in 1915,
several young girls reported seeing a headless angel-like
being hovering in the air.

A hunter in Nebraska saw a "magnificent flying
creature" emerge from a large cylinder that had landed,
like an airplane, near him in 1922. Several other people
from the area made the same reports.

In the USSR, a report in the winter of 1936 came
from the Pavlodar region of Kazakhstan. A fifteen-
year-old girl saw a flying man dressed entirely in black

above her as she walked to school one day. She said it made a "rumbling noise," was wearing some kind of headgear like a helmet, and had no face.

In Chehalis, Washington, right after the New Year in 1948, a woman and some neighborhood children watched a man hovering vertically in the air above a barn. He was about 20 feet above the barn and appeared to have silver wings and a control board strapped to his chest. The contraption made a "whizzing noise." There were also sightings by several witnesses of similar creatures in Longview, Washington, later that same year.

A couple living by the sea in Brazil were taking a walk one evening in the early 1950s. Hovering in the air about 30 feet above them were two flying humanoids. The creatures landed, appeared to hide, and then flew away. Their actions convinced the walking couple that the creatures might have been mates.

In 1951 and 1952, soldiers at Camp Okubu in Japan reported flying men about 7 feet in length, with a wingspan of 7 feet. Three Marines near Da Nang, Vietnam, saw a naked winged woman in 1969. The female creature had huge "batlike wings," black skin, and radiated a greenish light. The men were close enough to hear the flap of her wings.

Several police officers in Guadalupe, Mexico, saw something in the early hours of a winter morning in January 2004. At first, they noticed a dark cloud hovering near the ground, writhing and twisting. Upon reaching the ground, the cloud turned into a woman with black hair, dressed all in black. She was able to tumble and fly through the air extremely fast. Her eyes were entirely black, and she had no eyelids.

In LaCrosse, Wisconsin, in late September 2006, a man and his son saw a 6-foot-tall flying man with wings. They described his face as "hideous" and "full of sharp teeth." The creature screamed loudly when it nearly collided with the truck being driven by the witnesses.

Possible theories of these flying "humans" include that they are angels, demons, aliens, or government experiments. I think the possibilities of time and interdimensional travelers need to be further explored as well.

The Awful

The awfulness of "the Awful" must be beyond compare, as this monster is said to be the basis for many horror tales penned by the storytelling master, H. P. Lovecraft.

In 1925, Lovecraft traveled to the towns of Richford and Berkshire, Vermont, to investigate sightings of "a very large Griffin-like creature with grayish wings that each spanned 10 feet." The "serpent-like tail equaled its wing length" and its "huge claws . . . could easily grip a milk can's girth."

The creature was first seen by several sawmill workers as they were walking across a bridge in the middle of town. The beast was sitting on a rooftop looking angrily down at the men. One man was so frightened that he had a heart attack. He recovered, but for weeks he would awaken in the middle of the night screaming.

Spotted often over the next few weeks by farmers and their wives, locals were absolutely petrified. The Awful would fly over fields in the daylight, land on roofs, and appear to watch people as they went about their daily chores.

One witness claimed the creature was holding a little crying baby as it flew 50 feet over a field. No infants were reported missing in the area, so it was assumed to be a baby animal.

The creature was rarely seen after the 1920s— until recently. In October 2006, H. P. Albarelli Jr. wrote in the *County Courier* about a respected citizen of Richford, Vermont, who had seen the Awful appear from "nowhere and plucked a huge black crow" out of the top of a pine tree. The incredulous witness went on to say the creature circled his house three times.

After Albarelli's article appeared, other witnesses reported sightings of the cryptid. One lady talked about a time when she was about ten years old; she and her friends saw the monster "sitting in a tree near the Trout River." The large winged beast watched them for several minutes. She said it had a strange beak and reminded her of a pterodactyl.

A dowser in the Richford area, Edith Green, told Albarelli that people have been nervous about the Awful for a very long time. Another older gentleman said that the creature has been seen often in the Gibou area of Montgomery for the past quarter-century, even very recently. The man also explained that locals felt "we don't bother it and it don't bother us, maybe with a few exceptions."

One longtime resident of East Richford claims that the Awful has been seen recently around the Slide Road area. He added that "you can usually hear the thing before you see it." It makes a low screaming sound. When it is close enough, you can hear the flapping of its huge wings.

Mr. Albarelli is in possession of a "petrified jaw-bone" from an Awful given to him by a logger in the Richford area. It is "stone hard" and "bears a number of very large teeth." At last report, he was attempting to have the jawbone examined by an expert from the University of Vermont.

Beast of Busco
(Giant Turtle)

Years ago, I thought I had befriended a dinner plate–
sized turtle I'd rescued on the highway. I guess he felt a
bit differently, because when I put my hand in the box
where I kept him to straighten out his living quarters,
he chomped down on the back of it and wouldn't let
go. It was a painfully long time until my friends and
family were able to help me. My turtle-loving daughter
was born a few years later. I'm grateful that the kind
of turtles she happens to favor are of the illustrated
variety!

Not too terribly far from my encounter with the
turtle, the townspeople of Churubusco, Indiana, cel-
ebrate their own encounter with a turtle they call Oscar.

The Beast of Busco was named after the original
owner of the farmland, Oscar Fulk. In 1898, Mr. Fulk
told his neighbors that a huge turtle lived in the lake on
his land. Evidently the giant turtle hadn't caused many
problems for Mr. Fulk.

Years later, after he sold the farm, the new owners
saw the creature but really didn't know what they were
seeing.

In 1948, Oscar reportedly tried to take some fishing poles away from some fellows fishing at Fulk Lake.
They described the turtle's back as being as big as a
"dining room table." He is reportedly 15 feet long, more
than 4 feet wide, and weighs several hundred pounds.

The farmer who owned Fulk Lake reported in 1949 that Oscar had first started dining on his chickens and then had progressed to his veal! This was starting to cost the farmer some money, and he called for help from the police. They attempted to haul the Beast of Busco out of the lake with four Clydesdale horses. The chains broke, and Oscar escaped.

The town began to bustle with professional hunters and divers visiting the farm to catch a glimpse or more of Oscar. Oscar outsmarted them all and continued to break  free of their traps. One idea was to lure Oscar out with a 200-pound female sea turtle. Oscar didn't seem interested. Of course, no one really knew if Oscar was actually an Oscarina!

As the months went by, the owner of the land decided to drain Fulk Lake (I think that was a little far to go). Oscar was nowhere to be seen.

Other giant freshwater turtles with 12- to 15-feet shells have been sighted (and/or caught) in TaKua Tung, Thailand; Hoan Kiem Lake, Vietnam; and, Lough Bray, Wicklow County, Ireland. The Carvana of Texas is said to live in the swampy parts of the state.

In Africa, there is the legend of *Ndendeki*, a giant turtle who lives in Lake Tele, Congo. The natives have seen it often and know that it eats dead organic matter at the bottom of the rivers and other fresh waters of the region.

Lake Minnetonka, Minnesota, is home to a very odd-looking turtle claimed to be 30 feet long! He is also supposed to have stripes and a lion's mane in all the colors of the rainbow! A thirty-year-old, 10-foot long sturgeon called Lou is said to swim in the waters of this huge lake as well.

I can't imagine a turtle the size of a Volkswagen bus! I'm sure I would be certain it was a floating rock of some kind and try to crawl up onto its shell. Evidently these cryptids have been around quite a while, as explorers have found shell fossils 15 feet in circumference!

Giant Rabbits of England

I adore Jimmy Stewart in *Harvey*, the movie about the giant invisible rabbit. Of course, it would be rather difficult to "sight" an invisible rabbit. But, the townspeople of Felton in northeast England had their hands full in 2006 with a giant rabbit and his voracious appetite!

The enormous, brown-and-black, floppy-eared fellow made many nighttime visits to the gardens of Felton residents, leaving ruined vegetables and giant footprints in his wake. There is an area in Felton where twelve local residents grow vegetables that they enter into summer competitions. The giant rabbit has been

taking massive bites out of the cabbages, turnips, and carrots. He's strong enough to pull turnips and onions right out of the ground! The locals say he's a very hungry bunny!

Jeff Smith, one of the twelve gardeners, was the first to spy the furry vegetarian, saying he was the size of "a monster, a brute of a thing, absolutely massive" with huge footprints. After his initial sighting, "Bigs Bunny," as the locals have started calling him, has been seen by several Feltonians (many hoping to catch him in the sights of their firearms). One lady saw the big beast "thundering" across her yard on a rainy spring morning.

As crafty as Bugs Bunny, the giant rabbit continued its garden terror without getting caught, although sharpshooters, gamekeepers armed with air rifles, and animal welfare workers armed with traps made gallant attempts to catch it. Smith says that it is no ordinary rabbit; "it is clever." He adds, "They never see it. There were big rabbits in the 1950s and 1960s before pesticides were introduced, but not like this."

A pet store in Worcester claims to have the world's largest bunny, Roberto, a 35-pound rabbit measuring nearly 4 feet in length. Many speculate that Bigs Bunny might be an escaped "pet." These giant bunny pets are known for being able to escape to freedom and to breed with native rabbits.

A spokesperson for the British Rabbit Council of Newark states that the rabbits can be aggressive. They will protect what they perceive as their own territory.

The Flemish giant rabbits, possibly the result of giant Patagonian rabbits brought by Dutch traders from Argentina in the sixteenth and seventeenth centuries (for their meat) and bred with the large rabbits of Flanders, often reach weights of more than 22 pounds. The origin of British giant rabbits is not certain, but they grow to a maximum of 15 pounds.

In Eberswalde, Germany, Karl Szmolinsky raises giant German gray rabbits, which can reach a maximum weight of 23 pounds. (*See* photo on *ChicagoTribune.com.*)

Giant chinchillas weigh approximately 16 pounds and were originally bred in Kansas City in 1921, by mating a Flemish giant with larger-than-average standard chinchillas.

Whatever happened to Bigs? Happily for the gardeners of Felton (although it made me feel a little heart tug), he was hit by an automobile on May 19, 2006. The driver saw "a massive rabbit dart out across the road in front. . . . It was abnormally large." The impact cracked the car's bumper.

Although relieved, one local mentioned, "It looks like he's been busy though. I've never seen so many baby rabbits."

Thunderbirds
(*Piasa*)

When you're a ten-year-old kid, summer is a time of adventures and escapades. Some ventures can even be a bit risky and dangerous. Kids are willing to take a few hazardous chances. Mark Twain wrote all about that. During my many childhood adventures and escapades, one thing that never entered my head was the possibility of being picked up by a giant bird and carried down the block.

Not too many miles from where I was exploring caves, playing pirates, and swimming in the Mississippi River, and not too many years later, in the town of Lawndale, Illinois, ten-year-old Marlon Lowe wasn't thinking of giant birds either.

Late summer 1977. A rousing game of hide-and-seek near Kickapoo Creek with friends, steaks on the grill (maybe hot dogs or burgers for the kids), a perfect July evening with family and friends. Ruth Lowe stood at the door of the house and called the kids home to eat at a few minutes after 8:00 p.m.

Marlon claims in recent interviews that he didn't see, hear, or smell anything at all. He was running around the side of the house when something "shot down and grabbed him." The monster used its talons to grasp his tank top straps.

Ruth saw her son in the claws of one of a pair of gigantic jet-black birds with long white-ringed necks and curled beaks, wingspans of more than 10 feet, and bodies close to 5 feet long. Thankfully, the beast dropped Marlon after 40 feet, and the avian pair flew away, although many sightings were reported throughout the area after the incident.

There were several neighborhood witnesses to the attempted kidnapping. Marlon's mother, chasing after her child, will never forget the sight of her son's feet dangling in the air, his little fists hitting at the huge black bird as it carried him in his talons while trying to peck at his helpless body.

Hundreds of locals in central Illinois, claimed to see giant birds throughout the late 1970s. There were also reports from Alaska and Texas during this time.

The American Indians believed that there were two types of thunderbirds. One kind was helpful; images of the benevolent birds often decorated American Indian objects and jewelry. The other type was considered to be dangerous, often snatching children and eating them. Legends tell of the beast destroying entire villages.

Legends and sightings of giant birds come from all over the world, including the phoenix of the prehistoric Southwest, the *Demaj* of Persia, the *Imgig* of Mesopotamia, the *Anka* of Arabia, and the *Rukh* of Madagascar.

The Anka was known to snatch children and take them away. According to Arabic writers from the Middle Ages, the beast was called "Anka-mogrel because it carries off what it seizes."

Marco Polo wrote about the Rukh, saying it was "able to carry off an elephant." Natives told him stories of the bird appearing during certain seasons of the year. It was said to have a wingspan of 50 feet. John Tradescant, a British traveler and florist from the seventeenth century, claimed to have a claw from the Rukh in his possession.

Records of giant birds date as far back as the beginning of written language, found on ancient tablets in Mesopotamian ruins. There were huge winged monsters that could "carry off an antelope in each talon."

In northern Russia, there is a giant bird called *Vekher*, "wind demon." In Asia, giant birds are known as *Simurgh* and *Garunda*. In Iran, there are stories of a bird so large it could keep the rain from the Earth. Norsemen called their giant birds *Hraesvalg*.

Men building a railroad tunnel in northeastern France in the winter of 1856 claimed to see a huge winged creature with "sharp teeth, long talons, and a wingspan of more than ten feet" crawl out of a broken boulder and then die "at their feet." A paleontology student identified it as a pterodactyl.

In 1673, the first explorers of the middle Mississippi River area, Jacques Marquette and Louis Joliet, saw giant petroglyphs of a 40- to 50-foot winged beast on cliff walls near the merging of the Mississippi and Illinois rivers. Marquette wrote in his diary, "Each was as large as a calf with horns like a deer, red eyes, a beard like a tiger's, a face like a man, the body covered with green, red and black scales, and a tail so long that it passed around the body, over the head and between the legs, ending like a fish's tail."

In 1836, Illinois author John Russell wrote about "gigantic flying monsters" called the *Piasa* that terrorized the Illini tribe for centuries and the "Bird That Eats Men," whose cave near the Illinois River was littered with human bones.

Skeletal evidence of giant birds from prehistoric times has been found in several places. The largest fossil was uncovered in 1972 in Texas. This bird had a wingspan of more than 40 feet.

A live specimen of a giant bird was sighted near Danville, Illinois, in 1972. On a sunny summer day, John Walker was dove hunting when he heard what sounded like an animal caught in a trap. Then he saw a huge bird soaring in the air. Its wings were square, and its outer feathers were "frightening."

In 1977, near Lake Shelbyville, Illinois, Chief John Huffer, stringer for the local CBS affiliate, saw two huge birds roosting in a tree near a railroad bridge. They were jet-black, with huge necks and heads like "old cracked leather." They were clacking to each other with their beaks as they took off and circled above, before flying off. Chief Huffer said there have been "many many sightings of these giant Cherokee Thunderbirds in central Illinois."

The peak years for sightings in Texas were in 1974 and 1975, when dozens were reported; there was a large reptilian bird spotted as early as 1873 near a farm in northeast Texas by several farmers.

In early 1976, a man was attacked by a giant bird in a backyard in Raymondville. The beast was as "big as a man and had huge red eyes, large claws, and a wingspan of 10 to 12 feet." The man's clothing was ripped, but he was released, and the bird flew away.

A few weeks later, near San Antonio, three school-teachers saw a giant bird flying low in the sky one morning as they were on their way to school. They said it had a wingspan of close to 20 feet.

Another sighting in Texas occurred in Brownsville in 1995, at 4:30 a.m. Spotted by some early morning newspaper delivery guys, the giant raven-black bird was perched on a telephone pole. Eight feet tall, with a 15- to 20-foot wingspan, the monster had "stooped-up shoulders" and a "curved beak."

"Super-sized birds" in southeastern Alaska were reported by the press in October 2002. Witnesses said the creatures were the "length of a small plane." A pilot saw a huge bird flying alongside his plane. It was the color of bronze, with a big hook on its beak and a wingspan of 14 feet.

In my research, I discovered many accounts of children or infants being picked up by birds throughout the United States and Canada during the past 130 years. Some of the children were badly injured or killed.

One particularly horrible story from the *Manitoba Daily Free Press* in 1886 told of a two-year-old child whose brain was devoured by the bird through a hole it made in the small skull.

In 1924, evidence was found in a lime quarry in Taung, South Africa, of a small child who had been killed by a 14-centimeter-long talon that had pierced his brain. There were other punctures and slots, ostensibly made by the beak while the bird was devouring the child's eyes and brain. Another child's skull was discovered by wildlife artist D.M. Henry in the 1970s in a nest in Zimbabwe, and there are reports of small children being chased by monstrous birds that rip open their faces, eat their eyes, and devour their organs.

Some of the accounts I have read attributed the incidents to large eagles or condors. But many witnesses claim that the creatures do not resemble any modern bird believed to be in existence.

Personally, I think I'd rather the culprits be "gigantic flying monsters" than the good ole American eagle.

The craziest story I found about giant birds took place in our American Southwest. In 1890, two cow-boys reported a giant flying bird in the southwestern desert. They claimed to have shot and killed the beast. According to the April 26, 1890, *Tombstone Epigraph*, the bird's wingspan was said to have been 160 feet, the body was 92 feet long and 50 inches around the middle. The head on this huge bird was 8 feet long! It did not have feathers, but its "wings" were "composed of a thick and nearly transparent membrane." The story claims that two horses hauled this giant bird for quite a distance. Those horses must have been incredible beasts themselves!

Jake and Other Alligator Men

When I was a kid, my family moved around these United States often, and we took a lot of cross-country road trips as family vacations. I remember being enter-tained and intrigued by all the roadside signs announc-ing fascinating, creepy exhibits. I'm almost certain we passed signboards advertising "Jake the Alligator Man."

Unlike most cryptids, Jake the Alligator Man was quite social in his heyday. Witnesses say that the scaly beast enjoyed frequenting whorehouses and clubs and entertaining at carnivals when he took a notion.

He is currently on display in Long Beach, Washington, at Marsh's Free Museum on Pacific Avenue. The Marshes acquired the odd creature's mummified body for $750 in 1967 from an antique dealer. Mr. Marsh inherited his love of odd collectibles from his father, who had moved to Long Beach in 1935. In 1944, on a family vacation to Florida, Marsh discovered that a souvenir shop could prosper and decided to do the same in Long Beach. The business is still in the family, although it has moved across the street.

There have been documented reports of alligator men from the southeastern region of the United States since the late 1700s. Weighing in the range of 200 pounds and standing more than 4 feet in height,

the tribes of reptilian men travel under cover of darkness, in groups of five or six. Using swamps and waterways, they drift from eastern Texas to the Everglades of Florida, and even up into the Carolinas and back again. The beasts enjoy raw meat but can survive on leaves, nuts, and fruits.

The creature has a head and torso that are humanoid, but the lower body is that of an alligator, tail included. The alligator men have scaly skin that has a burnt look to it, like charred meat. They have sharp pointed teeth, slits for eyes, and stumpy claws.

In South America, there is a creature called the *Hombre Caiman* ("Alligator Man"). Naturally, it appears to possess features of both man and alligator. Sightings are very common in the rural areas of Plato, Magdalena, Colombia, and the story is that the hybrid was once a lecherous fisherman who had been tricked by a river spirit.

There was an account in a national tabloid, the *World Weekly News*, on November 9, 1993, of an alligator man that had been found in a Florida swamp. The report indicated that the mummified creature was discovered alive, hissing and carrying on in a "grotesque" manner. The story told quite an adventurous tale of the alligator man's "escape from captivity," "attack on a man in Miami," and his subsequent delivery of a baby . . . his own!

Dr. Simon Shute analyzed the beast, finding that the cranium was the same size as a human's. Mr. Marsh of the Marsh's Free Museum was not too happy with the report. Who can blame him? The picture of the "alive" and "hissing" beast was an exact replica of the picture of Jake used on the postcards he sells in his store. A picture of the very same creature on display in his free museum. A creature very much dead.

Beast of Bray Road and Other Contemporary Werewolves

I love to travel down isolated two-lane roads, feeling a part of the tranquil, country scenes, breathing in the fresh air, and happy to be alive. To me, the journey is just as important as reaching my desired destination. I guess most people prefer to make good time when they travel. Sadly, that means the interstates are taking over, and my favorite little roads are disappearing little by little.

On one country road outside of Elkhorn, Wisconsin, called Bray Road, there have been sightings since 1936 of a werewolf-type beast, with a rash of reports beginning in the late 1980s. The animal is described as a hairy, wolflike animal that walks on its muscular hind legs. He has humanlike hands and feet. Extremely large, the wolfman ranges from 5 to 8 feet in height and weighs several hundred pounds. The beast is often accompanied by the putrid odor of decaying meat. He often stares at his witnesses and appears to be intelligent. The Beast of Bray Road has been called the most famous of American werewolves, with many reported and documented sightings.

Mark Schackelman worked as a night watchman at a convent near Jefferson, Wisconsin. One night in

1936, he encountered a werewolf-type creature clawing at a Native American burial mound. The creature ran off when Schackelman approached him. Another night, Schackelman saw the beast digging at the burial mound again. The wolfman stood up and looked at Schackelman. It was more than 6 feet tall and hairy, with a muzzle, long fangs, and pointed ears. It smelled bad. But in this encounter, unlike the others, the creature spoke. He growled the word *gadarrah*. He looked at Schackelman for a very long time before walking slowly away. Mr. Schackelman considered it significant that Gadara is mentioned in the Bible as the place where Jesus encountered a demon-possessed man.

Close to this same area, in 1964, Mr. Dennis Fewless saw a creature weighing "between four hundred and five hundred pounds," dark brown, hairy, and nearly 8 feet tall. It ran across the highway in front of his vehicle and jumped a fence. Mr. Fewless was terrified.

In 1972, a large hairy beast attempted to break into a rural Wisconsin farmhouse. Unsuccessful, it left only to return a few weeks later and severely scratch a horse. A footprint more than a foot long was left behind.

Kim Del Rio was seven years old in 1977, when she saw a giant animal in her neighborhood. It had "human fingers, bushy hair, big teeth, big hands, and it was nervous and twitchy," according to sources on the *MonsterQuest* TV show. Kim was hypnotized to help remember more details of the incident, including the creature's very obvious state of anxiety.

M. Kirschnik, an artist, was traveling through Elkhorn, Wisconsin, in 1981 when he saw something standing behind a fallen tree. The creature made eye contact with Kirschnik. Decades later, Kirschnik still creates artwork with images of the creature he saw that day.

One autumn night in 1989, a young woman, Lorianne Endrizzi, was driving home from work on Bray Road. Off to the side of the road, she saw what looked like a person bent over. She was just a few feet from the being when she realized the "person" had fangs, yellow eyes, pointed ears, and a "long and snouty" face, "like a wolf." The creature was powerfully built, with "rather strange" humanlike arms. There appeared to be claws at the ends of the fingers. When describing it later, she referred to the beast as a "freak of nature."

That same year, hunters in Kenosha could hear heavy walking in the woods near them. They then saw a man-sized creature walking on two legs, growling loudly.

Doristine Gipson was driving down Bray Road on Halloween night in 1991 when it seemed that her right front tire might have hit something. She pulled over to check it out. From the darkness at the side of the road, about 50 feet from her, she saw and heard a large, dark, muscular creature running toward her! Immediately jumping into her car, she attempted to drive off, when the creature leapt onto her trunk. The car surface was wet and slick, and the beast fell to the ground. Doristine returned to the site later in the evening and saw a large shape moving on the ground.

Heather Bowey, living in the area as a child, saw the creature in December 1989 when she was outside in her yard with some of her friends. They thought they were looking at a huge dog, until it stood up. She said the creature's back legs were oddly shaped and that it moved incredibly fast, running and leaping. It was covered in "silverfish-like-brownish" hair.

A dairy farmer reported seeing a huge, "strange-looking dog" with pointed ears, long hair, and a large chest, in his pasture in the autumn of 1989. The farmer, Scott Bray, later found large footprints in the area.

Mr. Russell Gest of Elkhorn saw a similar creature near Bray Road during this time. The huge beast stepped out of a copse of bushes about a block from where Mr. Gest was standing. Mr. Gest claimed the creature was covered in black-and-gray hair, had a wolflike head, a big neck, and wide shoulders.

Another dairy farmer from Elkhorn saw the beast sitting by the road eating something he was holding in his front paws on a very early morning in March 1990. An occurrence at this time, possibly related to the sightings, was the appearance of more than a dozen dead animals in a ditch on a road nearby. The humane officer from the nearby town of Delavan expressed the theory that they had been used in some sort of cult ritual. Evidently some of the animals had ropes tied around their legs, and their throats had been slit. Many of the animals were missing their heads and other body parts. One dog's chest had been opened up and its heart taken out. Quite a few of the animals were identified as missing pets from the area. During this time, there were also reports of unidentified people posing as humane officers and collecting stray dogs. One child, home alone, told of a stranger who tried to talk him into relinquishing his black Lab. There were also reports of occult drawings found at an abandoned house and cemetery close to Bray Road.

In late winter of 1992, Tammy Bray saw the same beast her husband had encountered three years previously a few miles from Bray Road when she was driving home from work around 10:30 p.m. The wolfman crossed the road in front of her. Her description matched all of the previous sightings. She did add that it walked "strong in front, more slouchy, sloppy-like in the rear." A few months after this encounter, several horses were found in their pasture with their throats sliced open.

In the Honey Creek area of Wisconsin, a family on their way home from a Friday night fish fry spotted something on the bridge in front of their vehicle. It turned toward the car and stared at the family before jumping off the bridge. The red-haired beast was more than 7 feet tall and appeared to weigh 600 or 700 pounds.

Don Young, a hunting guide in the Bray Road area, has seen the werewolf five times since 2002. He describes it as "7 feet tall with brown-black hair, human feet, and black eyes."

On a sunny day in 2003, a young man saw a 6-foot tall creature in the local cemetery, with one foot perched on a gravestone. It had "big hair, a pronounced brow, wide nose, and an apelike mouth."

Katie Zahn was traveling through the state park in Rock County in 2004 when she saw some creatures "drinking water from a creek as a human would." In 2006, a deer hunter spotted a large monster that appeared hunched over and bulky, with human-looking feet. It could leap 12 feet in only two or three steps.

Some cryptozoologists suggest that this werewolf creature might be related to the *shunka warak'in,* a wolflike beast said to live in the forests of the upper Midwest. Its Ioway Indian name means "carrying off dogs."

Linda Godfrey, investigator and author of several books about the Beast of Bray Road (she named the creature), was convinced of the sincerity of the witnesses when she talked to them during her investigation. They genuinely felt they had seen something very unusual, and they were very frightened about what they had observed.

One eye witness she talked to was a young girl who had been driven up a tree by the creature behind a barn on her family's farm near Bray Road. She said the beast spent nearly an hour trying to get up the tree to her.

There was also a witness who saw two of these creatures together. Another believed the creature was about as close as you could get "to seeing a real werewolf."

Most of the people who talked with Ms. Godfrey indicated that the creature seemed "wary of humans" and preferred to not be seen. Several of the witnesses believed the creature to be some kind of supernatural entity. Ms. Godfrey stated in an interview that since the release of her first book she has had several sightings that "seem to have a paranormal component such as sudden materialization, morphing shapes, or telepathic communication." The supernatural aspect has been confirmed by some Native American sources.

One very interesting account told to Ms. Godfrey came from a bookstore clerk in Madison. He swears that he saw a "wolf-headed, human form morph into an ape-headed form" under a streetlight on a residential road in the early morning. He was still very frightened when relating the tale to Ms. Godfrey.

When the History Channel investigated the Beast of Bray Road for its series *MonsterQuest*, they subjected all witnesses to lie detector tests. Everyone passed.

It has been several years since the fifty-plus sightings of the Beast of Bray Road. Although the interstates are encroaching upon country road territory, there still remain many lonely old roads in Wisconsin. And every now and then, even as far away as Milwaukee, someone catches a glimpse of the contemporary American werewolf, the Beast of Bray Road.

Goatman

I'd be sure I was seeing Satan himself if I met up with a beast that looked half-human and half-goat. The "cloven hooves" stigma aside, my few experiences with goats at petting zoos lead me to think that the animals have a slightly crazy look in their eyes. (Forgive me, goat fans.)

Sightings of the creature called Goatman began in 1957 in Prince George's County, Maryland, but they have spread over the years to include most North American states and also Canada.

This human/goat looks much like the nature god, Pan, with an upper half-man body (although horns protrude from the head) and the legs and hooves of a goat. The satyr is said to be covered in fur, stand about 7 feet tall, and weigh several hundred pounds. The beast has no odor but does emit a high-pitched squeal.

Satyrs, Greek Gods with goat-like characteristics, are associated with sex and physical pleasure. The Goatman of modern times is known to frighten young lovers as they are engaged in arduous embraces by jumping on the hood of their car and pummeling it with the axe some say he carries.

There is disagreement as to whether or not the beast has harmed humans, but it has attacked and beheaded many dogs and other pets. It is also responsible for damaging property and destroying livestock.

Ginger, a pet dog owned by April Edwards, disappeared in early November 1971. The dog was found several days later without a head. April and her friends reported that they had heard strange noises and saw a monster walking on its hind legs in their yard that night.

There is a legend that Goatman is the result of a government experiment with DNA gone wrong at the Beltsville, Maryland, Agricultural Research Center. Is the creature what's left of Dr. Stephen Fletcher, the doctor who allegedly performed the experiment? Alternatively, did the doctor confess to creating Goatman by crossing the DNA of a goat with that of his assistant, William Lottsford, in an attempt to help his comatose wife, Jenny? Or, as some believe, is Goatman truly a satyr from Greek mythology?

Giant Catfish

As a kid, growing up in a small town on the banks of the Mississippi River, I remember hearing fantastic tales of divers seeing catfish as big as a man under the old Mark Twain Bridge that connected Missouri with Illinois. The monstrous fish could do nothing but recline on the bottom of the river, snatching little fishes swimming within reach of its hungry mouth. In Twain's *Life on the Mississippi*, Clemens tells of seeing catfish weighing 250 pounds and more than 6 feet in length— very near where I grew up (and swam in the river) in Clemens's boyhood hometown! But then again, old Sam was known for his fish tales.

After becoming an adult and moving around the country, I learned that most rivers, lakes, oceans, dams, and bridges are associated with similar stories of divers seeing gigantic catfish as big as "a man," "a cow," "a car," and even "a bus." Some accounts include frightening episodes resulting in the loss of limbs and tiny tots. I've read and listened to stories of accounts and legends of giant catfish from every state in the Union, nearly every country, and most continents.

Although the European catfish (*Wels*) is considered to be the world's largest catfish, living for more than 100 years and growing to more than 100 pounds, I'm sure fishermen and divers from most other countries would laugh at that claim.

A missionary, Eugene P. Thomas, fished out a bottom dweller weighing more than 200 pounds from the Oubangui River in the Congo. Franc Filipic of Ljubljana, Slovenia, almost caught a huge catfish in 1998. In fact, his last words were said to be, "Now I've got him," before he was pulled under by the beast.

A giant catfish called the *Manguruyu*, 18 feet long and weighing 1,000 pounds, is said to live in the Paraguayan Chaco, South America. In Rio Negro, Brazil, in 1975, a five-year-old child in a canoe was attacked by a catfish. He and his cousin were following the men of the town, who were gathering up all the fish left when the waters receded. The catfish jumped out of the water and hit the side of the canoe. The child fell into the water. He was never found. One week later, a red-tailed catfish was discovered on shore, dead . . . apparently as the result of choking on a small boy.

The giant catfish of the Amazon, called *Manguruyu Paulicea lutkeni*, is blamed for eating alligators and swallowing toddlers. Amazonian locals report that they regularly catch giant red-tailed catfish weighing more than 50 pounds and often pull out catfish weighing 200 or 300 pounds from the deep pools in the Amazon River.

Some American Midwesterners grew up fishing for the catfish by "hand grabbing" or "noodling." One noodler from Oklahoma said that the fish clamps down on your arm and spins "like a sharpener on a pencil . . . peeling the hide plumb off." Another noodler, with more than 1,300 noodles under his belt, says that it is very tough fishing. "Some of those fish are just incredibly, incredibly vicious." Although catfish don't have fangs, they do have rows of teeth designed to let food in but not back out. Many fishermen don't like noodling, and it is illegal in most places.

Catfish bigger than 15 feet have been seen in Lake Eufaula, Oklahoma. Vague claims of the fish attacking people have also been reported in that area.

In 1999, a Blue Catfish weighing more than 100 pounds was caught in the Ohio River. According to Jan Harold Brunvand, in *American Folklore: An Encyclopedia*, there is a report of a giant catfish caught in the Ohio River near Caseyville, Kentucky, in the 1970s, with a human baby found inside the fish.

At Elephant Butte Reservoir in New Mexico, there are claims that divers repairing a dam wall saw several catfish that compared to a "Volkswagen bug with the hood open." One woman told a fishing guide, John Morlock, that she had talked to divers who said the fish were "the size of school buses" and that one of the divers refused to enter the water again.

Perhaps the first documented mention of giant catfish in America was from the explorers Louis Joliet and Jacques Marquette. After they left the Peoria, Illinois, Indian tribe, a mammoth fish struck the explorers' canoe with such intensity that Marquette was sure the vessel had been damaged. They had been warned of a demon "who would engulf them in the abyss where he dwelt."

Mike Smith of the *Daily Lobo* mentions a missionary who told the story of a man being tugged into the Ohio River by a catfish in 1780; the fish drowned him before eating him.

In Thailand, a giant catfish weighing 646 pounds was caught in 2005 in the Mekong River. As big as a "grizzly bear," and nearly 9 feet long, the creature wrestled with a team of fishermen for more than an hour before they could bring it in. The Mekong Giant Catfish (*Pangasianodon gigas*) is listed by the *Guinness Book of World Records* as the largest freshwater fish ever documented, capable of growing to a length of 10 feet.

Before concluding this list of impressive bottom dwellers, I would like to tell you about another kind of catfish found in the Amazon called *Candiru*. Although these catfish are very small, they are deadly; perhaps even more so than their giant cousins. The Candiru have the uncanny ability to "smell" chemicals found in their prey's urine emitted from the gill cavities. The Candiru is then able to quickly attach itself to the prey with its spines where it chews a hole in a major blood vessel and feeds . . . kind of a vampire catfish. It then removes itself and sinks back down to the depths to digest its meal.

Also known to attack humans and other animals, the Candiru is said to swim into any orifice, including the anus and the urethra. Because of the fish's spines, it can only be removed through surgery. There are documented attacks on humans. A traditional cure is said to be the combination of the Jagua plant and the Buitach apple, which is then inserted into the affected area.

I'm certain I would much rather take my chances with the giant catfish.

Moas (Giant Flightless Birds)

It's almost surrealistic to drive down a Midwestern country farm road and see emus running in the pastures. The lean red meat of the emu has become increasingly popular in recent years. Similar creatures, the moas, are giant flightless birds once found on the islands of New Zealand.

These magnificent creatures were more than 12 feet tall, covered in reddish-brown feathers, and weighed hundreds of pounds. The largest were the females of the species. The moas had large skeletons with small skulls and short bills. A mummified skeleton of a moa shows that the creature had feathers for insulation on its legs, indicating that this species of moa lived in mountainous regions. These wingless birds subsisted on plants, fruits, grass, and twigs. Said to be extinct for centuries, moas have been sighted on the islands as recently as 2008.

The eleven species of moa documented are members of the Ratite family, which also includes the ostrich of Africa, the emu of Australia, the kiwi of New Zealand, the cassowary of Australia and New Guinea, and the rhea of South America. The extinct elephant bird of Madagascar was also part of this group.

In the early 1830s, a trader on the North Island of New Zealand heard accounts of moa being sighted in remote areas. There were many other accounts of large birds through the mid- to late 1800s, especially on the South Island. One report included a confrontation between a sheepdog and a moa.

There were additional sightings in the 1930s and into the 1960s. In 1989, a pair of moas was observed on the South Island. In 1990, tracks and sightings were reported from the Arthur's Pass area of New Zealand.

In January 1993, three hikers at Craigieburn Range, south of Arthur's Pass, noticed a large bird watching them. They were able to snap some pictures of it and some wet footprints the creature had left on a rock. A Department of Conservation officer confirmed that the images looked very much like a moa known to have lived on the South Island. In 1994, a doctor hiking in the same area discovered some damage that would have been consistent with known moa feeding practices.

There are still reports of the giant birds being seen in unpopulated areas of the South Island. In January 2008, after decades of searching, moa researcher Rex Gilroy discovered evidence of what he believes to be a colony of the smaller scrub moa considered to have been extinct since around 1500. Gilroy is convinced the tracks and a recently used nest, found in the remote Urewera Ranges in the middle of the North Island, belong to the moa. Gilroy said it would be possible for the creatures to have lived in the remote region for "hundreds of years" without ever being seen.

In nearby Australia, giant flightless birds called *mihirungs* are believed by some to still exist in remote areas today. There have been recent reported sightings of both the birds and footprints, and immense viable eggs larger than an ostrich's. The mihirungs can grow up to 10 feet tall.

I don't think I'll be seeing any moas anytime soon, although I'm not ruling out the possibility in the future. But, I do think I'd like to try an emu burger. At last check, ground emu meat was selling for $9 a pound. Healthy food always seems to cost more!

Shadow People

Shadow people is a relatively new paranormal term, but I have been hearing it more often lately, as reported sightings have been increasing worldwide. The entities are also called shadow folk, shadow men, and shadow beings.

Evidently, Art Bell, former host of the late night talk radio show *Coast to Coast AM*, is responsible for the recent resurgence of the term. The phrase appeared many years earlier, also in radio, as the title of a 1953 drama on Chicago's WGN-AM *Hall of Fantasy* about "malignant entities born of the darkness."

Heidi Hollis, researcher of shadow people, has been a guest on *Coast to Coast AM* several times. Hollis believes that shadow people have always been here, and that they are a negative influence. Possibly pockets of negative psychic energy that have accumulated in areas of traumatic events, they gorge themselves on fear and sadness.

Some believe that shadow people are ghosts or spirits, seen in varying stages of their ghostly development by people with different degrees of receptive abilities; much like the reception of a television or radio, which can provide a clear picture or not, depending on atmospheric conditions.

Members of the International Ghost Hunters Society believe that shadow people could be ghosts that don't have the energy or ability to manifest themselves into a more favorable or familiar shape. The society claims to have more than two minutes of videotape from Shaniko, Oregon, of a large group of shadow people ghost children "dancing across the wall of the old abandoned school building basement."

Some shadow people may be human-oid in form, but without defined features or resemblance to actual people. Many times, the shadows appear as two-dimensional or diaphanous. They often move very quickly, in a disjointed manner, but have also been observed to move slowly and fluidly. Normally, the shadow people appear just at the edge of your vision, caught out of the corner of your eye, or as a wisp through a mirror. They are able to quickly disintegrate into a wall or mirror. Occasionally, reports describe glowing red or yellow eyes. They have been seen as child-sized and also very tall and wearing a hat.

The tall shadow man called "Hat Man" is likely to be evil, according to Hollis. Hat Man appears in a much more solid and clearly defined state, wearing "a fedora hat, trench coat and three-piece suit." Hollis suggests the possibility that these negative entities might be attempting to "recruit people to the dark side."

They are reported to have many diverse personalities, ranging from shy to aggressive, and a few witnesses hold the opinion that they are guardian angels present to warn of imminent danger, much like the Mothman of West Virginia. (*See* **Mothman of West Virginia**, page 143.) Most feel that shadow people are accompanied by a feeling of apprehension, and some observers get the impressions that they are the "essence of pure evil." There have been reports of people being chased, attacked, and even raped by menacing shadow people.

The shadow people are sometimes compared to the Raven Mocker, an evil witch known in Native American Cherokee mythology, said to sometimes appear as a shadowlike phantom and to steal souls. It has also been suggested that there is a link between the *Jinn* of Islamic belief, which are made by Allah from black smoke, and the shadow people.

Although the possibilities appear to be endless—half-formed ghosts, demons, interdimensional beings, blobs of negative thoughts, and more—it is clear that shadow people are becoming one of the most regularly reported types of phenomena observed in recent times.

Rather than attempting to draw these creatures to you, it might be prudent to drive them away instead. I have received several requests for help in this situation at my paranormal website, *bellaonline.com/site/paranormal*.

According to Hollis, the number-one effective defense is to let go of your fear. These creatures feed on your terror. Stand strong! Think positively. It is similar to thinking happy thoughts in order to fly, as in the story of Peter Pan. Call upon help from your higher power. Burning sage is very effective in removing negative influences; sage grows easily in your herb garden (and fresh sage makes all the difference in your Thanksgiving stuffing!).

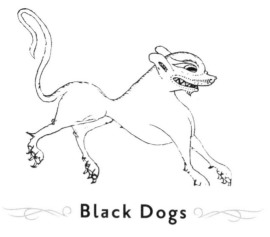

Black Dogs

One of the most intelligent, gentle creatures I have ever known in my life was a black Lab, given the very inappropriate name of Banana by my son at the age of six. This dog could open doors, sing for his supper, and even pack his belongings in the car when it was time to take a road trip (he didn't want to be left out)!

The black dogs I'm writing about are nothing like a beloved pet. Instead of love and adoration reflected in their eyes, there are red coals. Usually larger than a normal dog—even a big one—the creature sometimes seems solid, but at other times more ghostlike.

The beast's teeth are much larger than any dog's teeth should be. At times, there is a sulfuric or death scent attached to the evil mongrel. Occasionally, the dog is headless but, just as often, will howl eerily. Its paws can sometimes be heard hitting the ground. Other times, the dog is silent as a tomb. Black dogs are often sighted in graveyards and gallows sites, as well as roads and bridges. Often after the huge dog has been sighted, something terrible happens to the person, or family of the person, who has encountered the demon dog.

A happy little black terrier or spaniel, with sweet eyes and a wagging tail, greeted W. H. C. Pynchon, a geologist, as he made his way through the Hanging Hills of West Peak, Connecticut, in the late 1800s to study some unusual rock formations. Accompanying Pynchon silently all day long, the black dog disappeared on Pynchon's return leg as the day was nearing dark.

According to a story in *Connecticut Quarterly* in 1898, a few years after Pynchon's first visit to West Peak, he and a friend, Herbert Marshall, returned to the Hanging Hills together. They both confessed their similar "black dog" experiences from the area. Soon after their talk, the little fellow appeared above them, happily wagging his tail. As they made their way up to the creature, Pynchon's friend slipped and fell to his death. The legend connected with the sightings says that anyone who is visited once by the dog will have happiness; twice, sorrow; and if one sees the black dog three times, it means death. The appearance of the dog above Pynchon and his friend was number three for Marshall. Pynchon must have seen the black dog a third time a short time later, when he returned to West Peak. His body was found inches from where his friend had landed and died.

On Thanksgiving 1972, an alpine climber fell to his death in the same area. It is said he had recently been visited by the black dog for the third time.

The Native Americans have several legends that include black dogs and black ghost dogs being present when a catastrophe is on the horizon. There is a shamanic belief that the dogs guard the land of the underworld or "corpse ways."

I was amazed at the tremendous number of myths and folklore regarding black dogs, devil dogs, and hellhounds. They have been depicted as eaters of the dead and guardians of the road to hell. There is also a theory that they tend to be sighted at places located along the Earth's lines of energy.

The black dog has been seen for hundreds of years in the eastern coastal region of England, where it is called "Black Shuck or Scucca," sometimes "Doom Dog," "Barghest," "Hell Hound of Norfolk," "Old Snarleyow," "Mauthe Dhoog," and "Old Scarfe." Other names include Striker, Gytrash, Padfoot, and the Gurt Dog of Somerset, a benevolent dog who looked out for the children while they played and guided and protected solo travelers.

Black Shuck has an eye or eyes of red or green fire. Sometimes Shuck appears headless. He is usually surrounded by fog or mist and seen in out-of-the-way places such as graveyards, side roads, and deep woods.

One story from August 4, 1577, in Bungay, Suffolk, tells of a black dog appearing in a church one morning during a raging storm. Those who were touched by the creature died shortly thereafter. There is the image of a large black dog on the church weather vane to this day.

At another church only a few miles away, in Blythburgh, during the same violent storm, a black dog appeared, killing three people and leaving a burn mark, now called "the devil's fingerprints," on the church door.

In 1881, there was a sighting recorded about a magician who decided to see a Barghest for himself in Troller's Gill, Appletreewick. He called the beast to him, thinking that he was protected by the magic circle he had drawn around himself. He wasn't, and the black dog killed him. In 1890, in Norfolk, a boy claimed a large black dog chased him into the North Sea, from which he had to be rescued.

Only a few decades ago, in 1972, Nigel Lea, a traveler driving across the Channock Chase in England, saw a bright light fall from the sky. From the light emerged a huge, black dog with yellow eyes. After staring at the traveler for a few seconds, he disappeared into the trees.

By the 1980s, there had been enough sightings of the beast to earn it the name "Ghost Dog of Brereton." It was always described as large and black. Occasionally, it would just vanish into thin air.

In the year 1934, Ivan Vinnel was a young boy, living in nearby Burntwood. On their way home one evening, Ivan and a friend saw an unearthly "tall, dark man" with a black dog. They seemed to appear and disappear out of nowhere.

Years ago, in Bouley Bay, Trinity, an immense black dog with huge eyes was said to walk the cliffs, dragging a chain. At Hadleigh Castle in Essex, a black dog has been sighted several times since the 1970s.

In the folktales of the Missouri Ozarks, there is a story about an evil man who was lying in his bed, dying. Out of a cloudless sky, a bolt of lightning shot down and set the man's house on fire. Neighbors were unable to budge the man from his bed, nor were they able to move the bed itself. Just before the roof fell in, the neighbors escaped, as did a huge black dog from underneath the dying man's bed. An alternative version states that some locals killed the "witch" and his familiar, a black dog, with silver bullets and then burned them in a fire. No bones were ever found of either the man or the dog. There are sightings of a black dog in this area during traumatic events. Some hunters claimed the canine they saw was 8 feet long, headless, and that a thrown axe passed right through the creature.

The *Cu Sith* ("fairy dog") in Irish folklore is a large black dog with glowing eyes. As with many other black dog appearances, the Cu Sith is considered a portent of death and carries the human soul to his or her earned afterlife. The Cu Sith of Scottish mythology differs in that the hound is dark green. There is also a legend that the Cu Sith steals lactating women to feed the Celtic fairy children.

Church Grims are dogs that guard the interred of church cemeteries.

 In recent years, an angel of a dog that brings to mind our dear Banana has captured the hearts of the citizens in the West End of Richmond, Virginia. They call him Rasputin, and he's been watching out for the kids in that area for fifteen or twenty years. He's been seen to jump 6-foot-high fences without a running start; he has the ability to appear and disappear mysteriously; he loves to sleep in the snow or on an icy porch; and he even waits for green traffic lights.

Locals say he hasn't aged at all. There's no gray in his black coat. He always looks the same. He is said to have an "otherworldly aura," and his loving fans speculate about whether he is mortal, or even an alien!

As with most dogs, even the regular neighborhood canines, it isn't always easy to discern whether they have a sweet or nasty disposition. A mistake in that determination can result in a dog bite scar, similar to the one on my left thigh! If in doubt, it is sometimes more prudent to let "sleeping dogs lie."

Our sweet black dog, Banana, passed away a few days after my son left for boot camp in the Marine Corps. We miss Banana greatly. I have no doubt he'll be one of the four-legged creatures waiting to guide me to the other side when my time is here.

Kappa
(Japanese River Imps)

If you are ever walking by the river after dark and encounter a group of children playing in the moonlight, you might want to take a longer look. They might not be children at all, and they probably are not playing! You might be seeing a group of *kappa*, amphibious cryptids often mistaken for six- and seven-year-old children.

Kappa smell fishy, eat children and cucumbers, and are said to be very polite. They are the size of small children, with green, scaly skin, long hair, and sometimes carry a turtle shell on their backs. Some descriptions include webbed fingers. Their skin is wet and slick, although some are coated in a fur-like substance. Usually they walk upright, but sometimes they are seen crouching on all fours. Their heads have been described as monkey-like, but many times they have beaks like a tortoise.

Kappa have a concavity on the top of their heads that contains some kind of liquid, supplying the beast with power. If you chance upon one that doesn't appear to be too friendly, bow deeply. It will have no choice but to bow back, spilling its power-giving liquid and forcing it back to its water home.

Kappa are like chameleons in that they can change color. They appear to dislike loud noises and metal objects.

These creatures have been seen for hundreds of years. From 1600 to 1900, there was extensive study given to them. Several have been captured and documented. There are some vivid pictures in the 1820 compendium called *Suikokouryaku* (*see* Further Reading), which also identifies twelve types of kappa. Supposedly, in 1801, a kappa with a large chest, crooked back, and three anuses was netted in Mito, Japan.

One type of kappa, called the *Neneko*, was sketched in 1855 and is included in a volume entitled *Illustrated History of the Tone River.* Evidently the Neneko liked to move to different spots regularly, always causing havoc.

Most of the river imps were a little ornery, playing pranks and jokes. Some were evil, said to rape women, kidnap and eat small children or suck out their life force. One source I read indicated that signs warning about the kappa are still found near Japanese rivers and lakes in small villages.

Some kappa are friendly, even helpful with such farmers' tasks as irrigation. These kappa seem to have a high degree of integrity and will not break a promise.

There is a temple in Tokyo called Sougenji. According to the temple legend, the extensive drainage ditches in the area were built with the help of a kappa from the Sumida River. This benevolent kappa is said to bring business success to all who see it.

On a late night in early August 1894, a squid fisherman walking home from work in Tsushima encountered some children playing near the water. Unusual . . . so late at night. As he drew nearer, he saw their oddly shining skin. When he called to them, they quickly went into the river and were gone.

Next morning, the fisherman noticed "teardrop-shaped footprints" where the children had been gathered. The prints measured about 10 inches long by 5 inches wide and were 2 feet apart. The locals in the area seemed to have no doubt that they were from the river imps.

Local authorities took samples of the substance left in the prints, but the samples were small, and nothing was determined.

In late June 1991, in the town of Saito, an office worker, Mr. Matsumoto, and his wife returned home one evening to be greeted by an odor in their house. When they walked inside, they found small, wet footprints all over the house. Nothing appeared to have been taken or disturbed. There was an unidentifiable orange stain on an article of clothing. An identical stain was discovered on the stereo. Mr. Matsumoto collected samples to be analyzed. The orange liquid was found to be very high in iron and had a chemical composition "resembling spring water."

Mr. Matsumoto could not get over the invasion of his home; he consulted a shaman, who told him that it was the way of the kappa to play "occasional pranks" on the locals in the area, but that the kappa meant no harm. The footprints and orange stains proved impossible to remove.

Theories as to what these creatures could be include Japanese river deities, Portuguese monks who first appeared in Japan in the sixteenth century, or, horribly, "leech babies." There are stories from many years ago about impoverished people throwing their babies in the river because they could not afford to raise them. These unfortunate infants were unkindly called "leech babies." The Portuguese monk theory derived from the resemblance of monks' shaven pate to the depression in the top of the kappa's head. *Capa* was the Portuguese word for the monk's habit.

Many Japanese legends include tales of humans choosing to live happily with the kappa. There are also stories of kappa trying to drown horses and children in their rivers. More recently, a story surfaced of a small coffin with a mummified kappa found under an old floor in a sake manufacturing plant in Imari, Japan. My son plans to visit Japan one day. I am grateful that he doesn't care much for hanging out around bodies of water. As a major aficionado of video games, he won't be where the electricity is not!

Encantados (Dolphin-Men of Brazil)

One of the activities on my "list of things to do before I get too old" is to have the pleasure of swimming with a dolphin. In South America, there are shape-shifting dolphins called *Encantados* that can turn into humans at will. I think that would be a lovely creature to be. The name even sounds pleasing. Like an amphibious vehicle, one could walk on land as a human and then morph into a dolphin for travel through the waters of the Earth!

Encantado is a Portuguese word that means "enchanted" or "delighted" in English. In Brazil, it is also interpreted as a name for creatures who live beneath the Amazon River in a place called *Encante*. Many South Americans assert that, not only do they see and communicate with the Encantado, but that the creatures are their relatives!

Before my research, I had no idea that any freshwater dolphins existed at all! The *Tucuxi* is a freshwater dolphin living in the Amazon River. They are smaller than the long-nosed Encantado and resemble the bottle-nosed marine dolphin.

Also called *Botos*, Amazon River Dolphin, and Pink River Dolphin, Encantados often grow to a length of 9 feet. Their diet normally consists of crabs, catfish, and turtles.

According to legend, the Encantados love to love. They enjoy flirting and the physical act of love. The musical skills of the Encantados are excellent, and they have a deep love for parties with music and dancing.

Evidently, in Encante there is much happiness and wealth, but sometimes Botos desire the earthly pleasures and torments of the human world. Often, an Encantado falls in love with a human and will kidnap the object of its affection to take to the land of Encante. Many hybrid children have been born from these affairs.

It seems Encantados desire human company and will often capture children or other humans, who wander too close to the Amazon. Many Brazilians refuse to go near the water. There are stories that boaters have gone insane after an encounter with an Encantado on the river, although the creatures don't usually cause people any harm. I believe these people must have already had some mental instability or met up with an unusually evil type of Encantado. Some Botos have been said to inflict illness, insanity, and even death on humans. Perhaps these humans were deserving of such treatment.

Metamorphosis by a Boto into a human has been witnessed very infrequently, always at night. Once in a while, an Encantado stays too long at a party and will be seen running swiftly toward the river, as he changes from a human to his dolphin form. In human shape, Encantados appear very pale. They are said to move gracefully and enjoy dressing in colorful clothes. I believe they must use their powers to influence true humans to help them obtain clothing and other physical necessities. Always seen wearing a hat of some type, it is said that Botos retain their dolphin blowhole in human form. Sometimes, the transformation is incomplete, and one will see oddities such as Encantados with human hands at the ends of their flippers.

In addition to the ability to morph from dolphin to human and back, Encantados sometimes have power over storms. Some locals claim they have the ability to stun their prey with "bursts of sound" from their echolocation melon organ.

Similar to the siren's song or the vampire's stare, Encantados have the power to hypnotize humans. This is why many humans are irresistibly drawn to the river and taken to Encante. A special concoction of yucca root flour and dried crushed chili peppers, spread over the water at the location of the creature's appearance, is said to be effective in breaking the hold the Boto has over the human.

Parents in Brazil are especially wary at Carnival time, warning their daughters not to go off with charming strangers. It is considered dangerous to even make eye contact with an Encantado for very long, as that can cause nightmares for a lifetime.

Clurichauns

This little cryptid is one you might want to have around! Although they are mischievous, the clurichaun has some great attributes too! If you enjoy wine—and I do—you will want to pay special attention to the information about these little green fairies.

Clurichauns (Kloo ra kahns) are related to leprechauns and are generally considered to be the darker counterpart of the Irish fairy. A clurichaun is called a *monciello* in Italy and is also said to be called "His Nibs" in parts of Ireland. Its favorite pastime is drinking wine. They also enjoy riding sheep and dogs in the moonlight.

If you are friendly to the little fellows, they will protect your wine cellar at all costs! If you have a clu-

richaun around, and someone breaks in to steal from you, at least the wine is safe! Actually, most are very agreeable to protecting your entire home—it's their home too! If you happen to store your wine in casks, clurichauns will ensure that they do not ever leak.

Another huge benefit, in my mind, is that clurichauns like to be clean and well dressed. Most wear a red hat made from living organisms of some kind. Some are said to wear the colors red and white because they are offended if mistaken for their green-clothed cousins, the leprechauns. They don't usually carry any tools, as leprechauns sometimes do, because they usually don't have any desire to use them. Clurichauns are sometimes seen wearing blue knee-high stockings and silver buckles on their shoes.

If you don't treat these little guys (all those sighted have appeared to be male) benevolently, they can cause all kinds of problems, similar to poltergeists. They will even cut off their noses to spite their faces by spoiling the wine supply!

If you have a change of heart, decide you want one around after all, and attempt to rectify your unfriendly treatment, it will be too late. Once you drive a clurichaun out of your home, you can never have another.

Known to be happy little drunkards most of the time, they can be a bit surly before their first goblet of vino. A clurichaun prefers his own company to anyone else's and is quite content living a solitary lifestyle. Some people lucky enough to have one in their wine cellar say they are sometimes entertained by the little fellow singing Irish folksongs.

You can call a clurichaun to your home by leaving a little wine out for him after you go to bed. You might also try your own ritual or invocation to invite the little fairy to come to you. Be forewarned, though, you simply must have a supply of wine before you even attempt to call the clurichaun. And, you have to keep the supply well stocked! Be assured, if you treat your clurichaun right, he will be there to protect your fermented juice for your lifetime.

There's another possibility for bringing a clurichaun into your home that I hesitate to mention, because it can be extremely dangerous. You might try to find a circle of mushrooms called a fairy circle or fairy ring. According to folklore, these rings are doorways into the fairy kingdom. I have seen several in the past few years, about 6 feet in diameter, not too far from my home.

The largest and oldest known fairy ring is located in France. It is more than seven hundred years old and is about one-half mile in diameter. In England, on the South Downs, there are several large fairy rings many hundreds of years old.

Once you've found a circle, set aside a date on a night under the full moon. Be sure to take a friend with you to remain outside of the circle, in case you need to be pulled out. Sometimes the fairies don't like to let you go. Keep in mind that you might be invisible. Run around the circle nine times before entering. The fairies can be very helpful and entertaining, but they will want you to remain and dance with them, and they will attempt to lure you to stay with their beautiful music. Politely state your request for the services of a clurichaun, thank them, and then leave quickly, while you still have all of your faculties about you. A word of caution: if the fairies are not in a good mood, or you have found a group of less-than-friendly ones, you could be taken captive, put under a spell, and not released until you are very old, if then. A day in the fairy kingdom can be hundreds of years in human time.

Warning: There are some rogue clurichauns who have been known to steal from homeowners.

Giant Spiders

When my son was in Iraq, he told me of his encounter with a giant spider. It showed up at the door of his office.

He said it was huge, much bigger than a tarantula. The creepy arachnid looked at my son and then raised its front legs at him. Jimmy began throwing water bottles, the phone, the scanner, whatever he could quickly grab, while yelling at the beast, "Get away from me, you freaking demon!" The creature finally ran away, extremely fast. My son really wanted to come home then!

This must be one of those giant camel spiders I've been hearing so much about. Spiders that are said to be able to run 25 miles per hour, jump 6 feet in the air, have the ability to inject you with a numbing substance while you are sleeping so they can chew on a limb at their leisure, and lay their eggs in the bellies of camels.

Actually called *solifugids*, members of the class *Arachnida*, the eight-legged beasts are also called wind and sun scorpions. They are not really venomous, according to Rod Crawford, an arachnologist at the Burke Museum in Seattle. Some solifugid specialists disagree with this assertion and believe they do inject a type of venom that can paralyze their victims. These spiders are very strong, as well as fast, and can be found in the southwestern United States and southern Africa.

The solifugids are known as *matevenados* in Mexico, which means "deer killers," and the native people are very aware of how dangerous they can be. Crawford says their jaws are enormous so that "they can crunch their prey before it crunches them back." They have large fangs, called *chelicerae*.

If cornered, the solifugid will assume an aggressive stance, and it is able to run very fast, though only about 10 miles per hour rather than 25, as does the camel spider. They prefer shade, and they will dig holes to bury themselves during the heat of the day.

One witness described them as "something from a nightmare, with beady eyes, a hairy body, and jaws that bulged like Popeye's forearms."

Most giant spider stories come from the Congo of South Africa. Baka Pygmy Spiders are about a foot wide. Researcher Bill Gibbons wrote about the Lloyd family encountering a giant spider while they were "exploring the interior of the old Belgian Congo in 1938." They saw something on the track in front of them, which they thought to be "a large jungle cat or a monkey on all fours." They stopped their old Ford truck to allow the creature to pass by and were amazed to see that it was a huge brown spider. It looked like a tarantula "with a leg span of four or five feet." In South America, there is a spider called the Pinkfoot Goliath Tarantula, which has a leg span more than a foot long.

A tale of a 1948 sighting in Leesville, Louisiana, was reported by Todd Partain, researcher for *Cryptomundo*. William Slaydon and his wife, Pearl, were walking north on Highway 171 to church one evening with their three grandchildren. Slaydon heard a noise and quieted the children. As they watched in horror, "a huge spider, the size of a washtub" crawled out of a ditch. "It was hairy and black," and no one in the family was able to utter a word.

It walked across the highway and "disappeared into the brush." The family discontinued their evening walks.

One of the children present at the time, Richard, told the reporter that he always felt his "grandpa was familiar" with the creatures, and he always expected to hear about them again. But, he never did.

Many stories of unusual creatures have been reported by loggers over the years. William Slaydon worked in the logging industry, and it is possible that he had seen a similar creature previously in the deep woods.

Years ago, when my son was just a baby, we were driving on a country road outside of Salinas, California. Spying a huge spider (I am sure it was the size of a dinner plate), we stopped the car for a better look. The beast was very close to my car door, and I was looking out the window at it intently, when all of a sudden, it jumped into the air and I was looking at it outside my car window at eye level!

Boston Lemur

The Boston Lemur is a cryptid that came to public attention as I was working on this book. Because the creature was sighted by a well-known individual in sports, and because this may be the start of a string of sightings, I thought it would be beneficial to include the Boston Lemur in this field guide.

One night in midsummer of 2008, the Boston Red Sox senior adviser, Bill James, was walking home from Fenway Park. The crowds were gone, and he was alone on the street. The moon was full, illuminating the area fairly brightly. James noticed an animal that, at first glance, he took to be a cat. He immediately changed his mind when he noticed the creature had huge eyes on the sides of its speckled gray head. It also had a puglike face, a very long tail "like a broom handle," and moved in an "odd sashaying motion."

James stared at the animal for "probably six or seven times as long as the period that a fly ball is in the air," before it scuttled under a car parked by the road. At one point, it seemed to "lift its hind legs over a stick in the road by using its tail as a kind of lever."

As James continued his walk home, he thought about all the animals native to the area, but came up with nothing local. He finally decided it had to have been a lemur. Knowing full well that lemurs are primates native to Madagascar, and not to be found in the Boston area, he surmised that perhaps it was a zoo escapee. Against the advice of his wife (who was concerned about his reputation), James called the local animal control center.

He was informed that his was the first lemur sighting reported in the Boston area. All four lemurs in the local zoo were found to be present and accounted for.

There had been a previous lemur sighting 20 miles southwest of Boston on a rainy day in March 2002. On a farm in Sherborn, a "strange lemur-like dog" that looked like a smaller, "skinnier version of a Tasmanian wolf" was sighted by filmmaker Andrew Mudge.

Mudge was visiting his parents' horse farm in the rural area. Next to the farmhouse was an old, unused, junk-filled barn. Mudge walked out of the house and saw an odd creature hopping out of a hole in the door of the barn, heading toward the woods.

Mudge had grown up in the area, hiking all over the wilderness through his childhood years. He is very familiar with all the wildlife around there. He had never seen anything like this creature before. It was the size of a small coyote but looked like a combination of a lemur and a fox. It had a very long, hairless tail, except for a ball of hair at the end. The fur and face of the creature reminded Mudge of a lemur, with which he does have some experience.

Mudge has a degree in anthropology and studied lemurs in college. The creature he saw at his parents' farm had a faint striped pattern on its back, and the little beast had short ears and a long face.

Mudge followed the odd creature for a short time into the woods. It didn't appear to try very hard to get away from Mudge, but it didn't let him get too close either. As it got darker, and visibility diminished, Mudge returned home. Mudge's father did get a glance at the creature when Mudge first saw it and hollered to him, but not a close enough or long enough look to be able to describe it.

Bill James theorizes that the lemur could possibly have migrated in the few years between the Sherborn and Boston sightings.

Dragons

Australia is a country I hope to have the opportunity to visit someday, for many reasons. One is the wide variety of unusual animals found on the continent, including sightings of the *Megalania prisca,* a dragonlike creature thought to be extinct since the Ice Ages.

Some of the most immense creatures known to have existed in Australia before the Ice Ages were called megafauna. The largest of the meat-eating megafauna were the reptilians, such as the giant goanna, megalania, which often grew to 30 feet in length, and weighed more than 1,000 pounds, twice the size of its cousin, the Komodo dragon.

Large lizards were considered to be extinct until 1912, when a pearl fisherman reported seeing "enormous, prehistoric creatures" on the island of Komodo, part of the Lesser Sundas in Indonesia. A retired U.S. Army major claimed to have seen a giant goanna (or monitor lizard) when he was a young man, in 1913, at Emerald Creek in Queensland.

Several expeditions were sent to the island, but interest wasn't aroused until 1926, when W. Douglas Burden of the American Museum of Natural History went to investigate the claims.

Measuring 10 feet in length, weighing close to 200 pounds, and capable of killing water buffaloes, the giant lizards were called Komodo dragons. Known to run as fast as a man (even faster traveling uphill), this dragon's claws are large and curved, and its teeth are monstrous. They are known for their intelligence, which appears to be reflected in their eyes.

The dragons are still very much a part of Komodo Island today and have been discovered on other nearby islands. Although relatively rare, they have been known to attack and eat humans. The only item remaining of a tourist from Switzerland was his camera.

Komodo dragons have a very keen sense of smell, especially for rotting meat, which they will happily consume for their supper. If necessary, they will attack their victims by hiding and waiting until the prey is alone and vulnerable. Then the dragon will rush out and take a huge bite. The victim usually becomes unconscious from loss of blood or from the toxic bacteria in the dragon's saliva. The dragon's prey can include most animals, and they have been known to cannibalize, although their normal diet consists of deer or boar. They are said to be able to consume up to 80 percent of their own weight in one meal.

Recent sightings of the megalania, in Australia and elsewhere, are causing many experts to rethink the "extinct" label for this creature as well. During the 1960s, a French priest was traveling toward his mission, heading upriver with a native guide. The priest spotted a giant lizard sunning itself on a fallen tree trunk of equal length and, asked the guide to stop the boat. The native must have thought that was a crazy idea, because he refused. When the priest returned and measured the tree at a later time, it was 40 feet long!

In the summer of 1963, a 15-foot-long giant lizard was reported by the Karlsens, a couple traveling on a bush road between Brisbane and the Gold Coast. It came out of a ravine and ran across the road in front of their vehicle.

Four teenagers claimed to see a giant reptile on a jungle track ahead of them near Townsville, Australia in 1977. They described it as having a huge head, long neck, "enormous legs and big claws, and a long, thick tail." They said its body was "almost elephant-like," and covered with "large scales of a mottled grey colour." Its length was more than 40 feet, and it stood at least 6 feet on all fours.

In the summer of 1979, Australian cryptozoologist Rex Gilroy was informed of some huge tracks in a recently plowed field. Upon investigating, Gilroy discovered more than thirty tracks that appeared to be from a giant lizard. Although mostly ruined by rain, Gilroy was able to make a plaster cast of one; he was amazed by its resemblance to what might be left by a megalania.

That same year, herpetologist Frank Gordon was walking back to his vehicle after some fieldwork in the Wattagan Mountains of New South Wales. He saw what looked to be a fallen tree near his car, only to be startled when the 30-foot "log" suddenly ran off! A similar incident was reported by a surveyor in the area.

A farmer in the area observed a giant lizard approximately 25 feet long, while he was walking by one of his fields. A group of woodcutters reported a lizard they estimated to be 20 feet long.

In 1981, a soldier from Queensland reported an interesting sighting from October 1968, when he was with his unit on a jungle training exercise, on the Normanby Range. The unit came upon the remains of a dead cow in a swampy area. The bovine had literally been torn in two. They discovered "large reptile tracks and tail marks" in the area of the dead cow. It appeared to have been killed elsewhere, and dragged to the swamp. The unit left quickly.

Residents of Queensland report that large goannas live in the forests of Kuranda. Sometimes the giant lizards take chickens and calves. There have also been reported sightings by Aborigines from Cape York and Gulf country forests.

After speaking with so many eyewitnesses, and extensive investigation, Rex Gilroy is certain that these contemporary dragons, called megalania, are living in Australia today, and that eventually a live one will be found, much as its cousin, the Komodo dragon, was discovered in 1912.

Spring Heeled Jack

Victorian England isn't known for being an exciting era. Although, I'm sure for those who had the "pleasure" of meeting Spring Heeled Jack, it did have its moments!

The first reported sighting of this interesting character was by a businessman on his way home from work in 1837 London. He was startled while walking by a cemetery, when out of nowhere, a character jumped over the high railing and landed right in front of him!

The odd-looking personage before him was muscular and tall, with clawed hands and glowing eyes. He had a devil-like look to him with pointed ears, a sharp nose, and a goatee. He wore a black coat and a tight-fitting "oilskin" garment beneath, and he also had on a helmet. "Jack" was gentleman-like and, most alarmingly, perhaps, able to leap in huge, high bounds. He could jump over fences that were more than 9 feet tall, and he had an eerie laugh that would make your blood run cold.

Some have claimed that Spring Heeled Jack was able to speak English conversationally. He was also able to breathe blue-and-white flames and had claws with metallic tips.

After this, sightings of Jack began to increase, and he was also seen in Sheffield, Liverpool, the Midlands, and Scotland. Several females reported that Spring Heeled Jack touched them, even tearing at their clothes with his claws. Young Mary Stevens claims he kissed her face. Upon encountering the peculiar creature, some young women began having fits, and some were said to have died of fright!

Teenager Jane Alsop was accosted by Jack when he came to her door claiming to be a police officer. He tore her dress, but took off when people came to her aid.

A true police officer, James Lea, arrested Thomas Millbank in a pub called the Morgan's Arms not long after the attack on Ms. Alsop. Thomas had been bragging that he was actually Spring Heeled Jack. Millbank was wearing identical clothes to those described by

Ms. Alsop, but he escaped conviction because Jane Alsop insisted that her attacker could breathe fire. Millbank couldn't be proven to do this.

In the summer of August 1877, a group of soldiers in Aldershot's Barracks saw an odd figure bounding down the road, making a metallic sound. The character leaped over their heads and vanished. Then it reappeared and delivered several slaps to the face of one soldier with a frigid hand; evidently it had lost its claws by this time! Again, it disappeared.

In 1904, sightings of Spring Heeled Jack began to move in a westward direction. In September, in north Liverpool, he appeared on the rooftop of St. Francis Xavier's Church. He jumped off the roof and landed behind a house. Laughing maniacally, he rushed toward the crowd of people who were gaping at the sight of him. He leaped over their heads and disappeared.

In late spring 1953, thousands of miles away in Houston, Texas, several witnesses saw a character resembling Spring Heeled Jack perched in a pecan tree near an apartment building.

Near the Welsh border, in 1986, a salesman claimed to have encountered Spring Heeled Jack. The creature cavorted in huge leaps and bounds and had a devilish cast to his facial features. The salesman also said that the figure slapped his face.

Many theories have been offered to explain Spring Heeled Jack. Could he be an alien, a manifestation of the devil, or maybe even several different manifestations throughout history? One thing is certain: he was never caught. So remember when you are walking alone late at night and something huge sweeps through the air above your head: it could be a bat, an owl, or even Spring Heeled Jack. Unless you don't mind a slap or a kiss from a stranger, I'd get a little spring in my heels . . . if I were you.

Beast of Bladenboro, North Carolina

Another possible vampire cryptid, called the Beast of Bladenboro, or the Beast of Bolivia, North Carolina, was allegedly responsible for the deaths of hundreds of animals in 1954. Recently, reports of pets and livestock being killed in a 200-mile range of Bladenboro have incited people to start wondering if the monster they dealt with back then might be back.

More than fifty years ago, in Bladenboro, North Carolina, authorities discovered animals that had been savagely attacked. After finding unusual footprints, a large posse of men with hunting dogs and floodlights tracked a beast through the nearby woods.

The beast was first sighted on an evening in January 1954. A young woman walked out on her porch and saw a huge beast about 20 feet away. It was brown and looked to weigh about 150 pounds. The next day, large paw prints were found all over the property.

The sightings lasted for ten days, then stopped. "Crazy noises" accompanied the visitations, including squeals, "catlike" roars, and what sounded like a baby crying.

Parents would not let their children outside after dark. Residents of nearby towns refused to visit Bladenboro. Men wouldn't leave their homes without firearms. Dogs were locked up inside buildings for safety.

In recent years, animals have again been found

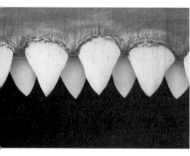

with their throats ripped open and their blood removed. Something nearly 5 feet tall, with a catlike face and vampire teeth, has been sighted in the area. Victims appear to struggle very little when attacked.

In September 2007, sixty goats were found murdered and drained of blood in Lexington, North Carolina. Glenda and Bruce Floyd, owners of the 100-acre farm where the goats were attacked, found them dead early one morning. Their necks had been ripped open and their bodies drained of blood, but they were left uneaten with no other marks. Thirty miles northeast of Lexington, in Greensboro, Billy Yow's goats were massacred in the same way.

In October, Bolivia, North Carolina, resident Bill Robinson found his three-year-old pit bull gutted. Robinson buried his pet "a fair distance away," but the next morning he discovered the carcass in the same spot where it had been the day before.

Four days after Bill found his dog's body, Leon Williams, a neighbor, found his two-year-old pit bull dead, with his chain "stretched out across the ditch." He was covered in blood, and pieces of him were missing. There were no signs of a struggle. The remains were examined by a vet, and it was determined that the dog had died of blood loss and a possible head dislocation.

The dogs were two of ten canines found killed in the areas of Brown, Midway, and Rutland Roads in Brunswick County during a two-week period. The tracks around their bodies, 4½ inches in diameter, were very similar to those of 1954. Recently, Robert Smith, a resident living on Brown Road, has been finding tracks around the church and in neighborhood gardens. He commented that he has never before seen tracks like these. "My fist could fit in it," he said. The animal services department in the county investigated the predator's tracks and droppings, but never conclusively determined what had attacked the dogs.

At the Buddhist monastery in the area, Abott Phra Vidhuradhammaporn was awakened early one morning by a loud sound that he was unable to identify. Because it was nighttime, he "did not want to go out."

Animal behaviorist Kay Cox told *MonsterQuest* (in their episode on vampire beasts) that she knows of no animal that could silently take down a 120-pound pit bull. The beast, whatever it may be, evidently does not alert the dogs to its approach. What animal kills for a reason other than food or fear? Is it killing to mark its territory?

Bladenboro, North Carolina, has made the beast business a lucrative one, with their annual Beastfest every year in October. If you're interested in tracking this cryptid, Beastfest may be a good place to start. Researchers and cryptozoologists often speak at the festival.

Ohio Grassman

When I was in third grade, my younger sister and I visited a carnival in West Palm Beach, Florida. One very memorable experience was entering the tent to see a woman turn into a gorilla. I will never forget the metamorphosis and still do not understand how it was done. When she broke her chains and started snarling and moving closer to us, we were out of that tent faster than we had ever moved in our young lives.

When I hear the story of the Grassman of Ohio, I imagine it looking like that gorilla woman. I'm not certain that I'm too keen on encountering this cryptid, but if you are, this is what you need to be looking for in the farmlands of eastern Ohio: The monster could weigh as much as 1,000 pounds, stands 7 to 10 feet tall, and has long flowing black or brown hair around his broad shoulders. His arms are muscular, his head is small, and he has no neck. The Grassman's hands and feet are very large. He has a wild odor, although it is sometimes described as smelling like rotten eggs. He has been known to make growling noises and other sounds.

The Grassman's diet consists of deer and farm food. He's often sighted near cornfields and other agricultural areas. The Grassman lives in a family-type unit; mothers and babies have been observed together. Sometimes the Grassmen are seen in groups of up to five.

When the Europeans first came into the area around Ohio, they saw the beasts in the grasses and assumed they were natives. They began calling them "Grassmen," according to Christopher Murphy, author and researcher. Murphy tells us that the Grassman was used as a type of bogeyman; parents used to tell their children, "Don't go out into the yard, or the grassman will get you."

On January 23, 1896, a man and his daughter were walking down a road near their home in Gallia County when a large beast attacked the man. After a frighteningly long struggle, the girl hit the creature in the head with a rock, and it ran away. They described it as being "gigantic in size," hairy, with burning eyes.

In 1988, truck driver Rich Lamonica was traveling near Salt Fork State Park when he spotted a black, wide, huge creature. He was able to see the beast clearly because of the snow-covered landscape and the small amount of foliage. The creature ran off into the woods.

Outside Akron, in 1995, researcher and head of the Ohio Center for Bigfoot Studies, Joedy Cook, found a primate nest of interwoven branches large enough for three adult men to fit inside. He attempted to investigate the nest further a few weeks later, but it was gone.

Late one night in Pleasant City, Ohio, in 1996, a woman was awakened by a barking dog. When she opened her eyes, there was something at her window and it began to growl at her. It backed away and walked toward her shed, turning to look back at her once. She described it as being very large, with "wideset eyes, a wide nose, and a big mouth." It walked on two legs. In 1997, in the wildlife area near Kershocktin, a tall two-legged creature was videotaped walking toward the woods.

Cook receives about a dozen eyewitness reports of the Grassman every year. In 2002, he received a cast of a handprint from a hunter in Beddenville, Ohio. A primate fingerprint expert examined the cast and found it to be from a nonhuman primate. In 2006, Cook found an odd skull and sent it to a primatologist in New York, who determined that the skull belonged to a male baboon from Africa. How on earth did the tracks and skull get to Ohio? Still, this can't account for the Grassman, who stands much taller than a baboon and walks on two legs.

In late December 2006, Treba and Terry Johns were enjoying the sounds and sights of the night at their campsite. It was almost midnight when they decided to pack up and leave. They saw a "broad shouldered, tall, black thing come up over the ridge." Treba went on to say, "Every nerve in my body was twitchin'. Every tiny hair on my face stood up."

When the *MonsterQuest* team investigated the Ohio Grassman recently, they heard sounds in the woods that included wood knocking, a form of primate communication. They also recorded vocal sounds of creatures that were not associated with any animals in the area.

I have no doubt the Grassman is a cryptid you could easily encounter if you choose to visit eastern Ohio. As anyone in the area would likely tell you, camp out in the woods near a farm or cornfield, and listen for the sounds of wood knocking and growling. I'd be prepared to defend myself . . . just in case. This fellow is really big!

Lizard Man
of Scape Ore, South Carolina

When I lived in the desert of Las Vegas, our family pet was a lizard named Rocky. This little guy was so cool. He would perch on my knee while I tried to get lost for a while in the adventures and mysteries of my books. We would catch crickets for him to eat. I cried when he died, and we had a funeral for the small reptile. You might be thinking I was a child when this happened, but I was actually a mother of two little ones!

In the Scape Ore swamps of Bishopville, located in Lee County, South Carolina, a 7-foot-tall bipedal creature with well-defined muscles, green scaly skin, and orange glowing eyes was frequently spotted in the summer of 1988. Additional features included three-finger hands ending in 4-inch black fingernails, and three-toed feet with circular pads that would stick to flat surfaces.

The first encounter with the Lizard Man took place about 2:00 a.m. on June 29, 1988. Teenager Chris Davis was driving home from work when a tire blew. He was just finishing up the tire change when he heard a "thumping noise" from behind the car. He turned

around to see the monster about 25 yards away, running directly toward him. "It looked strong and angry" as it attempted to tear the car door out of Davis's hands, then jumped on the roof while

Davis was trying to drive away. Davis swerved from side to side on the road, eventually throwing the beast off. His car roof had scratch marks, and his side mirror was damaged. Plaster casts taken of the tracks, spaced 4 feet apart, were labeled as "unclassifiable."

Tourists and hunters descended on the small town, and radio station WCOS offered $1 million to anyone who could capture the beast alive. Sightings declined by summer's end and, for decades, the Lizard Man was relatively quiet. Locals in Lee County, South Carolina, now fear the beast is back.

A resident of Newberry, South Carolina, reported two lizard men outside her home in October 2005. A vehicle was attacked by a lizard man in February 2008 in Bishopville. Dixie Rawson awakened one morning to find her van "chewed up." Pieces of the vehicle had been torn off and were all over her driveway. The parts were covered in blood and what looked to be "bite marks." The wheel wells on both sides, and other metal parts, were bent "like pieces of paper." Samples of the blood were taken for testing, but they were said to be contaminated, and no determination was ever made.

Dixie's husband, Bob, told interviewers, "I couldn't believe it, I just couldn't believe it. He literally bit, you can feel where he bit straight through here." He has his Glock loaded and ready.

Dixie believes there is something dangerous dwelling in the area; it is "very scary" living there now. Several of her cats have disappeared, and Dixie hopes they were just frightened off. Not long after the incident at the Rawson's home, several large farm animals were found dead nearby.

Yet, before the 1988 sightings of the Lizard Man of Scape Ore, South Carolina, were reported, there had been encounters with similar creatures in other North American regions.

The first report of a vehicle being attacked by a lizard creature was in November 1958. Charles Wetzel was driving near the Santa Ana River near Riverside, California, when a 6-foot-tall, glowing-eyed creature with "leaf-like scales and a protrusible beaklike mouth" attacked his car. Wetzel told author Loren Coleman, in a 1982 interview, that the creature's legs appeared to stick out "from the sides of the torso, not from the bottom." Wetzel made his escape by stomping on the accelerator, ejecting the monster from the hood of his car.

In August 1972, a lizard-type man was sighted in the Thetis Lake area of British Columbia by several eyewitnesses. The man-sized beast was humanoid in shape, with scale-covered silver skin, large ears, a monster face, and several "projections" on its head. A few years later, in 1977, a state conservationist reported seeing a "scale-covered man-beast" at dusk each evening near the waters of Southern Tier in New York State.

In 1982, a scientific paper was published by paleontologists Dr. Dale A. Russell and Dr. R. Seguin from the National Museum of Natural Sciences in Ottawa. The scientists theorized that, if the dinosaurs had not become extinct, a bipedal "dinosaurian counter-part of human beings" would have developed with three fingers on each hand. The model constructed from their description looks amazingly like the lizard man sighted in Scape Ore, South Carolina, today!

Lobo (Wolf) Girl of Texas

When I was in junior high school, I read a book that said people born on Christmas Day turned into werewolves when the moon was full. One of my best childhood friends, Holly (naturally) had been born on Christmas Day. She was a bit offended when I asked her about it on the school bus. Of course, I was with her many times when the moon was a full huge ball in the sky, and, although she could be quite ornery, I never did see her transform into anything even remotely wolf-like. However, a baby girl born in the year 1835 in the Beaver Lake area of Texas actually did make a transformation into a wolf girl.

In the spring of 1835, John and Mollie Dent had moved from Georgia to Beaver Lake so that John could do some trapping. One story relates that Dent was also hiding out after the murder of a colleague. Mollie was with child and began to have problems when it was the baby's time to be born.

John left to find help at a nearby goat ranch on the Pecos Canyon that was owned by a couple from

Mexico, but he was struck by lightning and died before he and the couple could make it back to Mollie. By the time the Mexican couple found the Dent cabin the next morning, Mollie had died. The baby was nowhere in sight. Some feared it had been eaten by a Lobo wolf, because of the tracks that were seen in the area and the fang marks on Mollie's body.

Ten years later, a young boy living near San Felipe Springs saw a naked girl traveling with a pack of Lobo wolves and attacking a herd of goats. A few months later, a witness saw two large wolves and an unclothed girl eating a dead goat in San Felipe. The wolves and the girl ran away. The girl was traveling on all fours, and then on two legs, running next to the wolves.

The Apache Indians told stories of small footprints and handprints seen with wolf tracks in the sand by the river. The wolf girl was hunted and found by the Apaches near Espantosa Lake, with her wolf friends, three days later. When she was separated from her companions, she fought mightily, biting, scratching, and making noises that sounded like something between a screaming woman and a howling wolf.

A male wolf came to her rescue and tried to fight her captors. He was shot and died immediately. The wolf girl fainted and was tied up and taken to a shack at a desert ranch, where she was locked inside. She refused covering, food, and water, preferring to curl up in a far corner of the shack.

She was said to be covered in hair, but she appeared to be human. She was physically fit, with extremely strong hands and arms. She was unable to speak, only growling. Her mobility appeared smoother and less awkward on all fours than when she was standing up straight.

She howled continuously, especially as night approached, and the calls were answered by wolves all over the area. Soon after, her wolf family came to her rescue. They attacked the animals at the ranch, and during all the havoc, Lobo Girl was able to escape.

There are only a few vague accounts of the wolf girl being sighted again, until 1852, when some surveyors, attempting to make a better road to El Paso, saw a wolf girl nursing two infant wolf cubs on a sandbar in the river. When she realized she was being observed, she ran away, carrying the cubs. She was not seen again.

There have been reports of "human-faced" wolves in the area for decades. In 1937, author L. D. Bertillion commented that he had seen "more than one wolf face strongly marked with human characteristics." There have also been reports from hunters of a white wolf girl apparition seen in the area.

One further note: Soon after moving from Georgia to Texas, Mollie Dent wrote her mother a letter that said only:

Dear Mother,

The Devil has a river in Texas that is all his own and it is made only for those who are grown.

Yours with love
Mollie

Giant Rats of New York City

I remember, as a kid, hearing about giants rats in the tunnels under New York City. I was happy that I didn't have to worry about huge rodents crawling out of the sewers and attacking me on my way to school. Although, given how much I hated school, that might have been a good excuse to not go!

According to experts in the city's pest control businesses and the homeless people who live in the streets and tunnels of New York City, the rats in the area are getting progressively bigger and more aggressive, even vicious. Numerous rats, the size of full-grown cats, have been spotted frequently under and above the streets of the city. They are becoming resistant to the poisons currently in use. The rats living on the Marshall Islands in the Central Pacific during the nuclear bomb testing programs went underground and did very well.

There have been giant rat sightings reported from other parts of the world, too. There were also prehistoric rats that grew as big as cars, with incisors more than a foot long, according to fossils found 65 miles west of Montevideo, Uruguay. Recently, in New Guinea, a giant rat five times normal size was found in the jungles.

Near the west coast of Madagascar, Giant Rats, the *Malagasy*, more than a foot long—and two feet long including the tail—and weighing about 3 pounds, live in underground tunnels of their own making. These rats are not aggressive and prefer to stay away from mankind.

The African Gambian pouch rat doesn't bring a rat to my mind as much as a squirrel. It can grow to weigh about 9 pounds and has cheek pouches for gathering nuts. These are friendly rodents and make good pets. Invasive in the Florida Keys, the pouch rats are believed to be responsible for the outbreak of "monkeypox." They were banned from the United States in 2003.

Put to use in Mozambique to sniff out land mines, the African Gambian pouch rats are nocturnal in nature and sometimes have trouble getting started in the morning! According to writer Michael Wines, rats are "outfitted in tiny harnesses and hitched to a ten-yard clothesline" to use their acute sense of smell to help their boss, Frank Weetjens, sniff out the treacherous mines. Although giants, they are still small enough to not detonate any of the mines.

The rats in New York City aren't friendly, aren't helpful, and are starting to scare many of the people in the area. In 1979, tugboat workers in New York went on strike, preventing barges from taking the trash off Manhattan Island. Rat problems got to be so terrible that the feral rodents would attack in the middle of the day. One witness described a lady being attacked by six or seven rats jumping and climbing all over her. A man started hitting the rats with a newspaper to get them off of her. A big pit of rats was found nearby, and it took weeks to clean it out, with the rats being carted away by the truckful. Thirty years later, the same alleyway is littered with the corpses of dead rats that appear to have been chewed on by other rats.

A maintenance man in the New York City area sees giant rats outside his apartment every night. He is terrified. Now he goes out with a hockey stick "every day to kill rats before they kill his children."

Several homeless men living under the city report that the rats are getting bigger and meaner. One homeless man, Jose, says he never brings food into the tunnels for fear of attracting the rats. He uses ammonia to keep the rats away from where he sleeps. There are some sections of the tunnels that he will never venture into. In a cave in one of the tunnels, he heard a terrible growling noise that still scares him to think about.

Lower Manhattan has more complaints than other areas, probably because the neighborhoods are older. Gino, a building manager in the city, was called by one of his tenants complaining about a rat in the toilet. He said the thing was gigantic, and he killed it with a baseball bat.

When I was living in West Palm Beach as a child, my best friend, Margaret Tibbs, called me one Saturday morning to tell me she had discovered a rat in her toilet that morning, when it tickled her rump. She really didn't seem as upset as I thought she should have been. The entire town would have heard my reaction to a rat tickling my behind!

It looks like giant rats are another type of cryptid that would be easy to encounter. Frankly, if I'm going to spend the money for a visit to New York City, it's not the giant rodents I'll be looking to see!

Dover Demon

Cryptozoologist Loren Coleman was the primary investigator of this cryptid and is responsible for naming it the "Dover Demon."

Sighted in the area of Dover, Massachusetts, several times in the spring of 1977, this creature is described as hairless with "rough, flesh-toned skin." The only facial features discerned were orange eyes, "like glass marbles," and no ears were detected. The demon's head is abnormally large, and it has long, skinny extremities and fingers that "grasp onto the pavement" and rocks.

On April 21, 1977, three teenagers were driving north on Farm Street, a road surrounded by woods and fields, in the Dover area. The driver, Bill Bartlett, saw "a bizarre, unearthly looking creature, close to four foot tall, crawling along a stone wall." Bartlett described it as looking like a baby with long arms and legs. He went on to add that its head was the same size as its body and was melon-shaped. What freaked him out most was that the creature turned his head to look at him as he drove by.

Bartlett's companions saw that their friend was very upset when he stopped the car a few moments later and reported what he had seen. They convinced him to drive back so they could try to get another look at it, but the creature was gone. Soon after the sighting, Bartlett drew sketches of what he had seen. He added this addendum to his sketch: "I, Bill Bartlett, swear on a stack of Bibles that I saw this creature."

A short time later that night, actually just a little after midnight, another teenage witness, John Baxter, was walking home from his girlfriend's house on Miller High Road when he saw a bipedal creature with a large head walking toward him. Not feeling any fright initially, John called to him, thinking it was another kid he knew who had suffered from a deformed head since childhood. There was no reply, but John and the small creature continued to walk toward each other. Finally when the two were about 20 feet apart, they both stopped and stared at each other. Then, the Dover Demon ran very rapidly into the woods at the side of the road. Baxter says he followed the creature for a while up hills and through ravines. He stopped near a creek, and he could see the creature above him on a hill, perched on a rock, and Baxter could "just barely discern the feet or whatever, you know, holding onto the rock sort of molding to the shape of the rock." The creature continued to stare at Baxter, making him very uncomfortable and scared. He backed up the bank and left the wooded area immediately.

The following day, two more teenagers, Abby Brabham and Will Traintor, reported a similar creature on the side of Springdale Avenue, but they described it as having glowing green eyes and being the size of a goat. This was the last sighting of the Dover Demon that spring.

Thirty years later, on April 22, 2007, the *Boston Globe* printed an article written by Kyle Allspice, who interviewed William Bartlett, now an accomplished painter in his forties. Bartlett still insists he saw the creature he described three decades ago.

Some cryptozoologists have made a connection between the Dover Demon and the *Mannegishi*, humanoid creatures from the Cree Indian culture. They are described as having long thin arms and legs, six-fingered hands, huge heads, and no noses. They live in the rocks and stones of creeks and rivers, enjoying the sport of canoe tipping.

The idea that the creature might be some type of alien is the most widely accepted theory in the field.

This creature has not been sighted in years, so looking for it in Dover, Massachusetts, might be a waste of time. Author W. Haden Blackman believes that the wilderness in Massachusetts, and into the forests of New Hampshire and Maine, would be more lucrative, especially areas near lakes and streams.

Considered one of the "Top Ten Most Mysterious Creatures of Modern Times," according to the *New York Times* website, *About.com.*, the Dover Demon has also been represented by toys manufactured by a Japanese toy company.

Mothman of West Virginia

Before seeing the movie, *The Mothman Prophecies*, in 2002, I, like most Americans, had not even heard of the creature. After I saw the movie, I became curious and did some research. It is still difficult for me to determine whether the Mothman was benevolent or malevolent.

In the Ohio River Valley, near the Chief Cornstalk hunting grounds, a father and daughter were amazed to see a huge man with wings fly into the air in 1961. Then, a couple of weeks after Halloween, on November 12, 1966, five men working in a local cemetery near Clendenin, West Virginia, saw a "brown human being" take off from a cluster of trees and fly over their heads.

On November 15, a man-sized beast with large mothlike wings and big glowing red eyes was reported in Point Pleasant, West Virginia, by the Scarberrys and the Mallettes. Driving by an abandoned World War II TNT factory near Point Pleasant, the two couples noticed "two red lights" in the darkness by the factory gate. When they stopped their vehicle, they saw the lights were "the glowing red eyes of a large animal, shaped like a man, but bigger, flesh-colored, with big wings folded against its back." Mrs. Scarberry remarked that its eyes dominated the head, and if you looked closely at them, they had a hypnotic effect.

Mrs. Scarberry went on to tell of the creature's wing being caught and it attempting to free itself with its "really big" hands. She believed the creature was very scared. When it managed to extricate the wing, it ran into the abandoned building.

A few minutes later, when the Scarberrys and the Mallettes were driving down Route 62 on their way back to town to tell the authorities, the creature began chasing them. Flying above the '57 Chevy at needle-burying speeds, making odd squeaking sounds, the Mothman pursued them all the way to the city limits, then flew off.

The next night, a posse combed the area, looking for the winged man. Two couples living near the TNT plant reported seeing the Mothman behind their parked car. It was in a recumbent position, before rising from the ground. Large and gray, with glowing red eyes, it watched them through their porch windows as they called the police.

A week later, Mothman was spotted flying over the region by four people. The next day after that, a witness reported seeing the creature standing in a field before "it spread its wings and flew alongside his car" until he reached the city limits.

One unusual sighting reported by Newell Partridge, a building contractor in Salem, West Virginia, included some odd details. He was watching television one night when the screen went dark. A "weird pattern filled the screen," and he

heard "loud, whining sounds from outside." Bandit, his canine companion, started barking and whining. The contractor walked outside to find Bandit near the barn. When Partridge focused his flashlight in that direction, he saw "two red circles that looked like bicycle reflectors." They were moving in the darkness and scared Partridge enough that he went back into the house and didn't come out until morning. Bandit was gone and was never seen again.

During November and December 1966, and all through the following year, there were more than a hundred sightings of Mothman reported from West Virginia. The creature was always described as having a 10-foot wingspan, large glowing red eyes, and provoked an accompanying feeling of dread. One witness dropped her infant baby when confronted by the Mothman.

Mary Hyre, a reporter for the *Messenger*, a newspaper based in Athens, Ohio, investigated the Mothman sightings. One weekend during the investigation, she received more than five hundred phone calls regarding "strange lights in the skies." On a night in January 1967, Mary was working late, when an odd little man walked into her office. "He was very short and had strange eyes that were covered with thick glasses. He also had long, black hair, cut like a bowl haircut." He had dark skin, looked "vaguely Oriental," and was wearing a black suit and tie. The little man seemed to have some type of

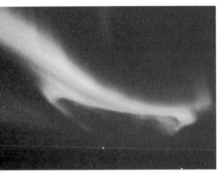

speech impediment; he asked about UFO sightings in the area. Ms. Hyre was very frightened of the man; "he kept getting closer and closer" to her and stared at her "almost hypnotically."

At one point, he picked up a pen from her desk and didn't seem to realize what the object was. Then, he "grabbed the pen, laughed loudly, and ran out of the office."

Reports from witnesses in the area indicate that the little man made several visits to homes whose owners had reported odd lights in the sky. He claimed to be a news reporter, and everyone he visited said he made them feel very uneasy.

A few weeks after that, when Hyre was on a street near her office, she saw the same man. When he noticed her watching him, he appeared to become distraught and jumped in a big black car that "suddenly came around the corner."

An increase in UFO sightings and "funny red lights" in the sky was reported in the area during the time of the Mothman, leading many to believe the creature might be an alien.

Point Pleasant is located between two wildlife management areas, and the empty TNT plant has "miles of subterranean tunnels" running under the buildings. What a perfect place for the creature to make its home while in the region!

After the Silver Bridge over the Ohio River collapsed due to a manufacturing flaw on December 15, 1967, killing forty-six people, the creature seems to have disappeared from the area. Many of the people who died had been principal Mothman witnesses, leading many to believe that the creature had been trying to warn of the danger. There are some who think it might even have been responsible for the disaster.

Although no documented sightings are on record after December 15, there were a few reports that officials seemed to "brush off" because they were dealing with the bridge catastrophe.

In December 1966, investigator and reporter John Keel had talked to many witnesses in the area and compiled information that included a number of poltergeist cases and other unexplained occurrences, such as cars stalling for no reason. Keel was certain all this activity was related to the Mothman sightings.

The Mothman was reported in Texas shortly after his disappearance from the Point Pleasant area, and I believe I might have had an encounter with the creature in New Mexico in the early 1980s. I was planning to explore an abandoned house about a half-hour north of Socorro one night with a friend. As we walked onto the porch and started through the front door, I glanced up and saw two large red eyes hovering in midair above me. Not easily frightened, and always ready for a ghostly investigation normally, I inexplicably became terrified! We ran to the vehicle and left immediately!

Later that night, after my friend and I had parted, I was driving home alone. I still felt more terrified than I had ever been in my life. Appropriately, John Fogerty's song, "Bad Moon Rising," began playing on the radio. I kept glancing in my rearview mirror, certain those red eyes were going to be looking back at me from my backseat. I arrived home safely. It was late, and I went right to bed. A little while later, I awakened to find my bed vibrating. That had never happened before, and I was still frightened from my earlier encounter. I looked around my room to see if anything else appeared to be vibrating, but my bed was the only piece of furniture moving in the room. I huddled against the headboard for an hour or so, too scared to reach over and turn on my bedside lamp, certain something was going to grab my arm. The bed continued to vibrate.

Suddenly, something tugged on my covers twice. Two very hard tugs. That did it! I jumped up so that I was standing on my bed. I reached over from that position and turned on the lamp, and then jumped off and away from the bed, running through the house, turning on lights as I ran. I found my Bible in my box of books and proceeded to read from it as I walked through the house. I interspersed words of warning with the Bible verses, telling whatever it was that this was my house and I wasn't leaving. As I did this, I began to feel less frightened and much stronger. Eventually, my fear lessened enough that I returned to the vibrating bed. Putting my Bible under my pillow and leaving the lights on, I finally fell asleep! The bed was still vibrating. The next morning, when I awakened, the bed had stopped moving, and whatever was causing the vibration did not return.

When I think back, I can't imagine how I was able to fall asleep. I honestly felt that I was stronger than whatever was causing this, and that it was important to show that I wasn't afraid. Monsters seem to feed on our fear.

New Jersey Devil

Although there are many devilish creatures throughout the world, there's a famous devil in New Jersey that is known as the Jersey Devil, also called the Leeds Devil and the Jersey Gargoyle. This particular horned one has been sighted thousands of times throughout New Jersey and in Pennsylvania over the past two hundred years, from the time when New Jersey was still a British colony up to the present.

The Jersey Devil has been compared to a monkey, a giraffe, and a kangaroo. It is usually reported to have a long neck, huge leathery wings, and split hooves. Some say it has a long tail, long legs, and a head shaped like that of a horse. It has also been described as having a head similar to that of a ram, with curled horns. Its height varies from a few feet to well over 8 feet. With glowing red eyes and an eerie high scream, it would be a daunting sight to encounter! The creature has been seen flying, jumping, and even squeezing through small spaces.

Many varying stories relate how the Jersey Devil came to exist in the Pine Barrens. One interesting legend tells of Deborah (or Jane) Smith (or Shrouds) from England immigrating to the Pine Barrens to marry Mr. Leeds. She bore twelve healthy children, but in 1735 she discovered that she was once again pregnant! This made her very unhappy, and she cursed the unborn child. The baby was born with cloven hooves, claws, and a tail (remarkably similar to the child in *Rosemary's Baby*). Mrs. Leeds cared for the strange child until her death. After that, the child took off for the Barrens, becoming the Jersey Devil.

Another version is that the newborn baby gargoyle ate his entire family immediately, then grew wings (as gargoyles tend to do) and flew up the chimney to reside in the Pine Barrens. There is also a story that a young woman fell in love with a British soldier, and her baby was cursed by the townspeople.

An interesting legend is that the creature was found by Abigaile (not Deborah or Jane) and Arthur Leeds and raised as their own. One similar tale is that Mrs. Leeds offended a preacher trying to convert her from the Quaker faith, and the priest subsequently cursed her forthcoming offspring.

Another popular folktale is that the bottomless Blue Hole located near Winslow, New Jersey, is actually a gateway to Hell, from which the Jersey Devil emerges at will. Interestingly, the Lenni Lenape Indians called the area now known as the Pine Barrens *Popuessing*, which means "place of the dragon."

There is a local legend that frightened citizens in the area and convinced a local preacher to exorcize the Devil in 1740. It is not believed to have worked indefinitely, as the beast has been seen countless times since.

In 1800, Stephen Decatur, a naval hero, saw the Devil while testing equipment in the Barrens. He is said to have shot and hit the creature, although it continued its flight through the sky.

Joseph Bonaparte, brother to Napoleon, is reported to have witnessed the creature while hunting on his estate in Bordentown, New Jersey, in about 1820. There are reports from 1840 of a strange beast killing livestock and scaring people in the area. Reports of sightings continued in 1873 through 1874 and started up again in the mid-1890s.

In 1900, the creature was seen by Mrs. Amanda Sutts on her family's farm. She had been ten years old when she saw the large animal. She said that it "sounded like a woman screaming in an awful lot of agony." The horses acted like they were going to tear the barn down.

During a week in January 1909, the beast was spotted repeatedly by thousands of people! E. P. Weeden, a city councilman of Trenton, New Jersey, was visited by the creature late one night. A police officer, a postmaster, and many business owners observed the winged oddity in Bristol.

Newspapers from all over the country followed the story and published numerous accounts by eyewitnesses. A lady hanging clothes in her yard swore that she saw it breathing fire. Mrs. Sorbinski in Camden chased it with a broom after it chomped a bite out of her dog.

There were numerous reports of cloven footprints, dead fish in the water, cows unable to produce milk, and attacks on pets and people during that time. Schools and businesses in the area closed for a period, because many of the towns were in such a state of panic. The sightings of 1909 ended with an attack on the creature with the town firemen's water hoses.

Some particularly entertaining stories of the Jersey Devil include reports of him dining with a Republican judge, frolicking with a mermaid in 1870, and enjoying evening walks with a headless pirate. The gargoyle has also been reported to be very "gentlemanly" in behavior toward the ladies.

In the early summer of 1926, two young boys saw what they claimed looked like a "flying lion." A cab driver reported sighting the anomaly in 1927. He was changing his tire outside of Salem, when the Jersey Devil jumped on his car. The driver jumped in and drove off.

In late summer of 1930, the Devil was seen by berry pickers eating blueberries and cranberries at Leeds Point and Mays Landing. It was reported by many witnesses throughout the 1930s and 1940s.

A few weeks before Thanksgiving 1951, a large group of children claimed to be cornered by the Jersey Devil at a clubhouse in Gibbstown. The creature left without hurting anyone. Near Morristown's National Historical Park, the creature was observed by four frightened witnesses in May 1966.

Two teenage boys spied a creature they described as the Jersey Devil while ice skating near Chatham in 1978. In 1993, the monster was sighted by a forest ranger in southern New Jersey. The two merely stared at each other before the Devil ran into the woods. Sightings of the Jersey Devil continue to occur, even as recently as 2008, with the most recent being in Litchfield, Pennsylvania.

My research indicates that in 1960, the Jersey Devil was witnessed by a large group of people, caus- ing the business owners of Camden, New Jersey, to offer a reward of $10,000 for its capture, which remains in effect. The Philadelphia Zoo is also said to have posted a million-dollar reward for the capture of the creature, which is said to remain in effect today. In comparison to some other cryptids, the Jersey Devil seems relatively harmless. . . . It might be worth looking into those rewards!

Chupacabra

I travel across the great state of Texas as least once a year on my way to see my son at Camp Pendleton. Last year, I saw something strange crossing a field and stopped my car to get out and get a better look. The creature and I stared at each other for what seemed to be several minutes before he continued on his way.

Although I thought of the recent Chupacabra sightings in the state, the creature was too far away for me to identify it. I just know that it appeared to have no hair, and its eyes were glowing. As I was writing this chapter, a news report came on Yahoo! News from Dewitt County, Texas. Sheriff Deputy Brandon Riedel caught an animal on videotape, in the daylight, running along in front of his vehicle. The creature had shorter front legs than back. When it turned its head to the side, you could see an incredibly long snout.

The *Chupacabra* (from Spanish *chupar* "to suck" and *cabra* "goat") was believed to be sighted first in Puerto Rico in 1995. Spotted thousands of times since then, it has been seen as far north as Maine and as far south as Chile. In 2006, the Chupacabra was sighted in central Russia, reportedly killing and draining the blood of dozens of turkeys and sheep.

Although the descriptions vary, what comes to mind when reading about the creature's pronounced eye sockets, humongous fangs, sharp claws, glowing slanted red eyes, forked tongue, and batlike wings, is the bloodsucking vampire. The Chupacabra has a row of spikes or quills down the length of its back. It is usually described as hairless with leathery or scaly spotted skin. This cryptid is said to be from 3 to 5½ feet tall and to travel extremely fast by hopping or running. It hisses and screeches when startled (similar to the sounds a vampire makes when exposed to sunlight) and has been said to leave behind a stench of sulfur.

Some accounts of the monster's description include immense dark wings that give it the ability to fly. Many witnesses compare it to an alien, or even a dinosaur, because its shape is similar to that of a "Grey," with its slit for a mouth and only tiny holes for a nose. Reports vary as to a tail. Some descriptions include an elongated jaw and a covering of hair on the body.

The attack of the Chupacabra is very rapid, with complete exsanguination of the victim through one, two, and sometimes three holes of various sizes. There have been instances reported of the victim's organs being sucked out through the holes. There is speculation that the beast injects some kind of toxin that totally immobilizes the victim before sucking out the blood. It is also theorized that the punctures are patterned in such a manner that death is instantaneous.

Other reports say that the marks are more like clearly defined bites, usually around the softer tissues of the body. A slimy substance is left behind at the wound.

In addition to chickens and sheep, the Chupacabra preys on cows, horses, cats, rabbits, turkeys, geese, and ducks. A favorite delicacy is goat meat.

The Chupacabra usually prefers to hunt at night, as bright lights are said to bother the beast. It has its brazen moments, though, as it has been seen walking in broad daylight. A young college student in Puerto Rico saw it tear apart the family goat on a sunny afternoon.

The Chupacabra was first reported and named in 1995, when it killed eight sheep in Puerto Rico. There wasn't a drop of blood left in any of the animals, and they each had three puncture wounds in their chests. More significantly, two fishermen claimed to have been chased by the creature. In November 1995, a woman said a Chupacabra stuck its clawed hand through her window, grabbed a teddy bear, and shredded it to pieces. Hundreds of attacks were reported over the next few months, and civil defense officials were called in to investigate the death of farm animals with "distinctive puncture wounds" on their bloodless bodies. *El Chupacabra* is blamed for the deaths of more than one thousand animals in a one-year period. The animals are always left with puncture wounds and no blood remaining in their bodies.

In 1995, Mr. Melendez saw something by the far railing on his balcony. He quickly went into the house, and the creature stood and watched him from outside for a short time. Melendez said it was unlike anything he'd ever seen. He had never been so scared in his life.

Researcher Mark Davenport and filmmaker Joe Palermo were in Puerto Rico during the time of the attacks. They reported that El Chupacabra was in the news there constantly. One neighborhood in Canovanas, Puerto Rico, was visited by the Chupacabra every night for four months!

In 1996, in the Sweetwater district of Miami, Florida, two ranchers lost a total of sixty-nine animals in a single night. The slaughter was witnessed by an elderly woman who talked about the incident publicly.

That same year, in Tucson, Arizona, the Espinoza family called 911 to report a Chupacabra in their home, on their seven-year-old son's bed! Another report came in from Tucson a few hours later. Both witnesses stated that the hairless creature "smelled like a wet dog."

In 2000, J. L. Talavera shot a Chupacabra in Nicaragua. Evidently, a veterinary scientist looked at the creature, was not able to identify it, and suggested it might be the consequence of genetic engineering. The creature was sighted again at various locations throughout the world, including Tucson, Arizona, in 2003; St. John, Indiana, in 2004; and, the Dominican Republic in 2005.

Throughout the summer and fall of 2004, several ranchers in Texas claimed their livestock were being killed by a creature fitting the description of the Chupacabra. In Elmendorf, Texas, in 2005, a rancher named McAnally reported something was stalking and killing his chickens. They were left completely bloodless. In one attack alone, thirty of his chickens were murdered. McAnally said he saw a strange creature four different times. The last time, it was 30 yards from him. He shot and killed it. The photos show a four-legged animal with big teeth and skin similar to an elephant's in color and texture.

Motocross racer Kolt Jarrett spotted an odd creature with "spikes down its back" and a "weird-shaped head" in November 2005, at the Cycle Ranch

Motocross Park in Floresville, Texas. Jarrett thinks he had an encounter with the Chupacabra.

Reggie Lagow, a farmer in Coleman, Texas, reported trapping a creature appearing to be "a mix between a hairless dog, a rat, and a kangaroo." The animal had been killing his livestock. He sent the beast to Texas Parks and Wildlife, but stated later that the "critter was caught on a Tuesday and thrown out in Thursday's trash."

Near Alamogordo, New Mexico, cattle roper Trey Rogers saw a Chupacabra in the forest while he was playing paintball in September 2006. It was reddish brown, with spikes down its back. Rogers said it appeared to have wings on its side. He said it was the "fastest thing" he'd ever seen.

Ben O'Quinn and his son, Tyrell, discovered a hairless creature under the porch of their home in Pollock, Texas, in 2005. It was smoky gray, with huge teeth and "crusty skin." Ben shot and killed the beast and took pictures of it.

"Odd roadkill" found in the Cuero area by Phyllis Canion in August 2007 was examined but was not identified. The canine teeth were described as unusually large, and the skin looked "unnatural." Canion also reported sightings of this creature on her ranch during the time her chickens were being killed. She said the "meat was left; it wasn't taken like with most animals. The blood was sucked out." She has the creature's head in the freezer at her home. Veterinarians and pathologists have examined the wounds left by the Chupacabra and report they were not caused by any known animal.

In the spring of 2006, Russian news reports began to include sightings of a Chupacabra and the finding of dead turkeys and sheep, with the blood completely drained. The dead remains of a creature described as a Chupacabra was found by a road near Turner, Maine, in the summer of 2006. There had been several reported sightings of the extraordinary beast, and unexplained dog killings, in the area for many years previously.

The following spring, in 2007, news reports from the area of Boyaca, Colombia, talked about massive numbers of sheep found dead. There have recently been reports of the Chupacabra killing chickens in the Philippines, Mexico, and California. Although most reported sightings are from recent years, there are stories of a creature similar to the Chupacabra from back in the 1950s in Arizona.

The origin of the Chupacabra is debated among

researchers and cryptozoologists. There is always the alien theory. That seems to pop up with most cryptids. Some Puerto Rican people believe that the creature might be the result of some type of genetic experiment conducted by the U.S. government in the El Yunque mountain area.

In some towns in Texas, festivals and celebrations are held in honor of the Chupacabra. On the other hand, local governments in Puerto Rico and Mexico tend to take the matter of the Chupacabra very seriously. They are very worried about the animal attacks, the impact on the ranchers, and they fear that their children might be next.

In San Paulo and Rio, Mexico, locals believe the Chupacabra steals children in the poor areas of town, draining their blood and leaving the bodies in dirty alleys. In Mexico City, there have been reports of the beast hunting schoolchildren, and in May 1996 a nurse reported that her arm had been severed by a Chupacabra. A farm boy from Jalisco claimed to have been attacked by the creature and had fang marks on his body.

This cryptid is considered to be danger-ous. An article entitled "Chupacabra Sightings in California" in the June 30, 1996, edition of the

Advocate Herald warned people in the San Diego area to call local authorities immediately if they encounter the creature. "Do not approach it and move indoors as quickly and quietly as possible." If you are bitten by the beast, get to the hospital as soon as possible, advises author and researcher W. Haden Blackman.

El Chupacabra has a very strong sense of smell. Chances are, if you are brave enough to hunt the mysterious beast, you will instead find yourself being stalked by the Chupacabra. My daughter has recently moved to New Mexico, and often takes solitary drives through the desert and mountains with only her dog as a companion. It is my sincere wish that she does not encounter El Chupacabra—at least not until I'm there with her!

A Tale of MoMo (Missouri Monster)

Recently I drove one of the most beautiful, scenic Mississippi River roads in the country, Highway 79, to Louisiana, Missouri, where I talked with longtime resident Gail Suddarth of Louisiana. In 1972, when Gail was about seven years old, she was instrumental in perpetuating the MoMo legend.

In my research of MoMo, I had read several accounts of the Suddarth story. As Gail talked, she revealed to me a very different tale of what happened that early August morning.

It was a lazy, late summer morning, and Gail was outside by herself picking tomatoes in the family garden. Gail was a little bored that morning and started thinking about the big monster scare that had been sweeping the area over the summer. A precocious little girl, she decided to have some fun. She had no idea how far that fun was going to go!

She had seen pictures of the MoMo footprint and had a pretty good idea of how to make one of those footprints herself. Very carefully, with her big toe, she drew out an enormous footprint in the soft dirt of the garden. She pressed her little foot down in certain places until it looked just the way she thought it should.

Then, Gail went running excitedly into the house to tell her family what she had "found" in the garden! Her mom and dad followed her outside to take a look. They were pretty impressed with their daughter's finding, all right! They called their friends, who called their friends, who called their friends until, before you could say "Missouri Monster," the media and the experts had descended upon the little girl's family farm!

What could she do? What had started out as a little fun with her family had turned into a huge ordeal! She bravely kept her secret all to herself. She didn't share it with anyone at all, not a sibling or a friend.

Tests were run, plaster casts were taken, and the ground was studied. "Oh my, yes," the experts agreed, "this was definitely a genuine footprint from MoMo." The pressure points were perfect, and the print was in the soft sand of the garden, which "explained" why there was only the one print. They listed all the convincing points verifying that this was indeed the footprint of MoMo, the Missouri Monster.

Before the swampy land of southeastern Missouri was drained to become the rich farmland of today, sightings of an immense, hairy humanoid responsible for killing livestock were reported often. After a hunter recounted shooting at the black hairy beast, it disappeared for nearly thirty years.

Appearing again in July 1968 in a St. Louis suburb, the hairy beast grabbed a four-year-old boy playing in his backyard. When the child's aunt started screaming and chasing the creature, it dropped the toddler and ran into the woods. Trackers found no evidence of the beast.

In 1971, Joan Mills and Mary Ryan had stopped for a picnic lunch off of Highway 79, when they smelled the most horrible odor anyone could imagine. Standing in the thicket was a half-ape and half-man creature. Ryan said "the face was definitely human," but "it had hair over the body as if it was an ape. It was more like a hairy human." The cryptid was covered in hair, except for the hands, which were hairless.

Mills stated that it made a gurgling sound, as it shambled toward the women, who were running to their car. The monster stroked their car and even tried to open the doors. Unable to leave because they had neglected to grab the car keys (just like in a typical horror movie scene), the ladies began honking the horn, which made the creature back off. But, before heading back into the woods, the hairy monster grabbed a peanut butter sandwich and devoured it in one bite!

Dubbed "MoMo" for Missouri Monster by the press, sightings were reported from all over the Louisiana area. In July 1972, eight-year-old Terry Harrison and his little brother Wally were playing by the woods next to their house. Their older sister Doris heard the boys scream. Looking out the window, she saw a monster "six or seven feet tall, black and hairy" standing "like a man," close to the boys. It had a dead dog in its arms and was covered in blood. The creature disappeared deeper into the woods, but the family dog became extremely ill for several hours afterward, possibly from the overpowering stench. Clumps of hair and several footprints were found in the area.

Neighbors began to report missing dogs and other large animals, footprints, handprints, beastly smells, and the sounds of an animal "carrying on something terrible." A group of teenagers claimed MoMo confronted them, roaring "ferociously." A driver told the story of the hairy monster attempting to overturn his car, with the driver inside!

Quite a few sightings were reported until the winter of 1972, when the creature might have gone  into hibernation in one of the numerous caves honeycombed throughout the hills in the area. People in the area of Louisiana, Missouri, still report seeing MoMo in the vicinity now and then, but the creature (and/or its kin) seems to have improved its hiding techniques. I have been in numerous caves throughout this region but have not yet discovered any evidence of MoMo, although, in the mid-1970s, in a cave located in Hannibal, Missouri, my friend Morgan and I did find eating utensils and other tools, indicating someone might be living there. I am very sensitive to putrid odors, though, and noticed no unusual smell at all. Thank goodness this was not the den of MoMo, because Morgan grabbed the flashlight and ran for her life, leaving me to feel my way back in the dark!

Gail Suddarth did not come forward with the truth for many years. She was especially reluctant to do that for a while, as her mom continued to see the creature in the yard, intermittently, throughout the years. She thinks her mom has finally forgiven her (but she still has that plaster cast)!

Further Reading

General

Blackman, W. Haden. *The Field Guide to North American Monsters.* New York: Three Rivers, 1998.

Coghlan, Ronan. *A Dictionary of Cryptozoology.* Bangor, ME: Xiphos, 2004.

———. *Further Cryptozoology.* Bangor, ME: Xiphos, 2007.

Coleman, Loren. *Mysterious America.* New York: Pocket Books, 2007.

Coleman, Loren, and Jerome Clark. *Cryptozoology A to Z: The Encyclopedia of Loch Monsters, Sasquatch, Chupacabras, and Other Authentic Mysteries of Nature.* New York: Simon & Schuster, 1999.

Cryptomundo (Cryptozoology museum), Portland, Maine. *http://www. cryptomundo.com*

Newton, Michael. *Encyclopedia of Cryptozoology: A Global Guide to Hidden Animals and Their Pursuers.* Jefferson, NC: McFarland & Company, 2005.

Shuker, Karl. *The Beasts That Hide from Man: Seeking the World's Last Undiscovered Animals.* New York: Paraview, 2003.

Rods (Skyfish)

http://www.americanmonsters.com/content.php?section=articles&idarticle =234

http://www.askoxford.com/concise_oed/cryptozoology?view=ukhttp:// www.bibleufo.com/ufos6.htm

http://www.burlingtonnews.net/ufooverburlington.htm

http://www.flyingrods.com/articlesfl/kfmb1.asp

http://www.meta-religion.com/paranormale/Rods/rods.htm

MonsterQuest Episode 111, "Fourth Dimension."

http://www.newanimal.org/air-rods.htm

http://www.paranormal.about.com/library/weekly/aa061598.htm

http://www.realsightings.com/rods.htm

http://www.rense.com/general50//whatthe.htm

http://www.roswellrods.com/story.html

http://www.subversiveelement.com/Rods.html

http://en.wikipedia.org/wiki/Rod_(cryptozoology)

http://www.xprojectmagazine.net/archives/cryptozoology/rods.html

Mongolian Death Worms

http://cryptoworld.co.uk/projects/operation deathworm

http://www.forteantimes.com/features/articles/158/death_worm.html

http://www.newanimal.org/deathworm.htm

http://www.virtuescience.com/mongolian death worm.html

http://en.wikipedia.org/wiki/Mongolian_Death_Worm

http://www.youtube.com/watch?v=c1hxIQYNdgQ<

Ahool and Other Giant Bats

http://www.americanmonsters.com/content.php?section=articles&idarticle
=225

http://www.americanmonsters.com/monsters/avian/index.php?detail=artic
le&idarticle=218

www.answers.com/topic/sasabonsam

http://creationwiki.org/Indava

http://www.fortunecity.com/roswell/siren/552/af_konga.html

http://www.fortunecity.com/roswell/siren/552/af_sasabonsam.htm

http://meta-religion.com/Paranormale/Cryptozoology/Other/ahool.htm

http://www.monstropedia.org/index.php?title=Sasabonsam

http://www.my-indonesia.info/page.php?ic=1123&id=3270

http://www.newanimal.org/gbats.htm

http://www.newanimal.org/orang bati.htm

http://pages.prodigy.net/jonwhitcomb/scpterosaur/

http://www.ppne.co.uk

http://www.ropens.com/indava

http://www.unknown-creatures.com/batsquatch.html

Whitcomb, Jonathan. *Searching for Ropens.* Livermore, CA: Wingspan Press, 2007.

http://en.wikipedia.org/wiki/Ahool

http://en.wikipedia.org/wiki/Kongamato

http://en.wikipedia.org/wiki/Olitiau

http://en.wikipedia.org/wiki/Ropen

Loogaroo Vampires of the West Indies

http://www.bbc.co.uk/dna/h2g2/A273566

http://dominica.dexia.dm/index.php?action=artikel&artlang=en&cat=30 &id=240

http://enchanteddoorway.tripod.com/vamp/trinidad.html

Maberry, Jonathan. *Vampire Universe.* New York: Citadel Press, 2006.

http://www.monstropedia.org/index.php?title=Asema

http://www.monstropedia.org/index.php?title=Loogaroo

Rose, Carol. *Giants, Monsters, and Dragons.* New York: Norton, 2000.

http://www.unicorngarden.com/vampires.htm#vampires

Bunyips (Water Horses)

http://www.americanmonsters.com/monsters/aquatic/index. php?detail=article&idarticle=11

Barrett, Charles. *The Bunyip, and Other Mythical Monsters and Legends.* Melbourne: Reed & Harris, 1946.

http://www.cryptozoology.com/cryptids/bunyip.php

http://www.dcn.davis.ca.us/~btcarrol/skeptic/bunyips.html

http://www.drizabone.com.au/legends/bunyip.html

http://www.newanimal.org/bunyip.htm

http://www.pantheon.org/articles/b/bunyip.html

Smith, Malcolm. *Bunyips and Bigfoots: In Search of Australia's Mystery Animals.* Alexandria, Australia: Millennium Books, 1996.

http://versaware.animalszone.lycos.com/continents/oceania.asp

http://en.wikipedia.org/wiki/Bunyip

http://en.wikipedia.org/wiki/Kelpie

Kentucky Goblins

Bloecher, Ted, and Isabel L. Davis. *Close Encounter at Kelly and Others of 1955.* Evanston, IL: Center for UFO Studies, 1978.

Carlton, Michele. "Children of Witness Defend Father's 1955 Claim." *Kentucky New Era,* December 30, 2002.

Clark, Jerome. *Unexplained! 347 Strange Sightings, Incredible Occurrences, and Puzzling Physical Phenomena.* Detroit: Visible Ink Press, 1999.

Coleman, Loren. *Mysterious America.* New York: Paraview, 2007.

Dorris, Joe. "Kelly Farmhouse Scene of Alleged Raid by Strange Crew Last Night: Reports Say Bullets Failed." *Kentucky New Era,* August 22, 1955.

———. "To Affect Visitors." *Kentucky New Era,* August 22, 1955.

Edwards, Frank. Lecture, "The Civilian Saucer Intelligence of New York," at the Pythian Temple, New York, NY, April 28, 1956.

Hendry, Allan. *Ronald Story: The Encyclopedia of UFOs.* Garden City, NJ: Doubleday & Company, Inc., 1980.

Schneiman, Sarah, and Pat Daniels. *Mysteries of the Unknown: The UFO Phenomenon.* Des Moines, IA: Time Life Books, 1987.

USAF Project Blue Book records, Project card 10073.

http://en.wikipedia.org/wiki/Kelly Hopkinsville_encounter

Tommyknockers of California

http://www.answers.com/topic/bucca 4

Butler, Lisa. "Miners Kept Safe With Help From Tiny, Magical Men." *County Times & Review,* August, 1999.

Calhoon, F. D. *Coolies, Kanakas & Cousin Jacks.* Sacramento, CA: Cal-Con Publishers, 1986.

http://www.legendsofamerica.com/GH Tommyknockers.html

Leifchild, John R. *Cornwall: Its Mines and Miners.* New York: Longman, Brown, Green, and Longmans, 1855.

http://www.monstropedia.org/index.php?title=Knocker

http://en.wikipedia.org/wiki/Tommyknocker

Enfield Horror (Giant Kangaroos)

Clark, Jerome, and Loren Coleman. "Swamp Slobs Invade Illinois." *Fate Magazine,* July, 1974.

Coleman, Loren. "Mystery Animals Invade Illinois." *Fate Magazine,* March, 1971.

http://www.cryptozoology.com/glossary/glossary_topic.php?id=424

http://www.prairieghosts.com/enfield.html

Taylor, Troy. *Haunted Illinois.* Decatur, IL: Whitechapel Productions Press, 1999.

Frogmen of Loveland

Clark, Jerome. *Unexplained!* Detroit: Visible Ink Press, 1999.

Coleman, Loren. *Mysterious America: The Revised Edition.* New York: Paraview Press, 2001.

http://www.ufocasebook.com/caponi.html

http://www.ufologie.net/htm/caponipics.htm

http://www.unknown-creatures.com/loveland-frog.html

http://www.unknownexplorers.com/lovelandfrogmen.php

Woodyard, Chris. *Haunted Ohio II: More Ghostly Tales from the Buckeye State.* Beavercreek, OH: Kestrel Publications, 1992.

http://www.xprojectmagazine.com/archives/cryptozoology/lovelandfrog.html

Gray Man of Ben MacDhui

http://www.bigfootencounters.com/creatures/greyman.htm

http://www.bigfootencounters.com/creatures/wudewasa.htm

http://www.book-of-thoth.com/thebook/index.php/Brenin_Llwyd

Cooper, Susan. *The Grey King*. New York: McElderry/Atheneum, 1975.

http://www.ghost-story.co.uk/stories/benmacdhui.html

http://www.monstropedia.org/index.php?title=Fear_Liath_More

http://www.newanimal.org/biggrayman.htm

http://www.pantheon.org/articles/f/fear_liath_more.html

http://www.spookystuff.co.uk/FearLiath.html

Trevelyan, Marie. "Water-Horses and Spirits of the Mists." In *Folk-Lore and Folk-Stories of Wales*. London: E. Stock, 1909. Available online at *http://www.red4.co.uk/ebooks/trevfolklore.htm*.

http://en.wikipedia.org/wiki/Fear_liath

Lusca and Other Giant Octopuses

http://www.americanmonsters.com/monsters/aquatic/index.php?detail=article&idarticle=101

Anon. "Bermuda Blob Remains Unidentified." *ISC Newsletter* 7, 1988.

———. "Giant Octopus Blamed for Deep Sea Fishing Disruptions." *ISC Newsletter*, 1989.

Benjamin, G. J. "Diving Into the Blue Holes of the Bahamas." *National Geographic*, September, 1970.

Clark, J. *Unexplained!* Detroit: Visible Ink Press, 1999.

Dinsdale, T. *Monster Hunt*. Washington, DC: Acropolis Press, 1979.

Ellis, R. *Monsters of the Sea*. New York: Alfred A. Knopf, 1994.

Heuvelmans, B. "Annotated Checklist of Apparently Unknown Animals with which Cryptozoology is Concerned." *Cryptozoology* 5, 1985.

———. *In the Wake of the Sea Serpents*. New York: Hill and Wang, 1968.

Mackal, R. P. *Searching for Hidden Animals*. Garden City, NJ: Doubleday and Co., 1980.

MonsterQuest Episode 211, "Boneless Horror."

http://www.naturalhistorymag.com/master.html?http://www.naturalhistorymag.com/editors_pick/1971_03_pick.html

http://www.nektoncruises.com/Destinations/Lusca.aspx

Pierce, S. K., G. N. Smith Jr., T. K. Maugel, and E. Clark. "On the Giant Octopus (Octopus giganteus) and the Bermuda Blob: Homage to A. E. Verrill." *Biological Bulletin 188*, 1995.

Raynal, M. "Properties of Collagen and the Nature of the Florida Monster." *Cryptozoology 6*, 1987.

http://www.strangemag.com/globsters1.html

Verrill, A. E. "The Florida Sea Monster." *American Naturalist 31*, 1897.

———. "A Gigantic Cephalopod on the Florida Coast." *American Journal of Science 3*, 1897.

———. "The Supposed Great Octopus of Florida; Certainly Not a Cephalopod." *American Journal of Science 3*, 1897.

http://en.wikipedia.org/wiki/Kraken

http://en.wikipedia.org/wiki/Lusca

Wood, F. G. "Stupefying Colossus of the Deep." *Natural History*, March, 1971.

Flying Man of Falls City, Nebraska, and Other Flying Humanoids

Clark, J. *Unexplained!* Detroit: Visible Ink Press, 1999.

http://paranormal.about.com/cs/humanenigmas/a/aa082503.htm

http://www.ufocasebook.com/unsolvedflyinghumanoids.html

http://www.unknown-creatures.com/flying humanoids.html

http://www.wormwoodchronicles.com/lab/flyingmen/text.htm

The Awful

Albarelli Jr., H. P. "Has the Awful Returned to Berkshire & Richford?" *The County Courier*, October 19, 2006.

Coghlan, Ronan. *Further Cryptozoology*. Bangor, ME: Xiphos Books, 2007.

www.cryptomundo.com/2006/10/page/2/

http://www.ghostlytalk.com/node/1000

http://www.ghostvillage.com/ghostcommunity/index.php?showtopic=17364

http://hsnl.org/phpBB2/viewtopic.php?t=364

www.paranormal.about.com/od/paranormalgeneralinfo/a/news_061024n.htm

Beast of Busco (Giant Turtles)

Campbell, Elizabeth Montgomery, and David Solomon. *The Search for Morag.* New York: Walker, 1973.

http://www.cryptozoology.net/english/africa/congo_republic_of_the/ndendeki/overview.html

http://www.newanimal.org/gturtle.htm

http://en.wikipedia.org/wiki/Beast_of_Busco

Giant Rabbits of England

http://www.chicagotribune.com/features/lifestyle/green/chi cc5rab bit20080821093148,0,5970604.photo

http://www.cryptomundo.com/breaking news/wor rabbit/

http://news.bbc.co.uk/1/hi/england/tyne/4886272.stm

http://news.nationalgeographic.com/news/2006/04/0411_060411_rabbit.html

http://www.slashfood.com/2006/04/10/monster rabbit terrorizing uk gardens/

http://www.timesonline.co.uk/tol/news/uk/article702852.ece

http://en.wikipedia.org/wiki/Flemish_Giant

Thunderbirds *(Piasa)*

AAA Illinois/Indiana/Ohio Tour Book. Heathrow, FL: AAA Publishing, 1997.

AAA Michigan/Wisconsin Tour Book. Heathrow, FL: AAA Publishing, 1997.

Austin, H. Russell. *The Wisconsin Story.* Milwaukee, WI: The Milwaukee Journal, 1948.

Balesi, Charles J. *The Time of the French in the Heart of North America: 1673–1818.* Chicago: Alliance Francaise, 1991.

Berlitz, Charles. *World of Strange Phenomena.* New York: Fawcett Crest, 1988.

———. *World of the Odd and the Awesome.* New York: Fawcett Crest, 1988.

Blashfield, Jean F. *Awesome Almanac—Illinois.* Fontana, WI: B&B Publishing, Inc., 1993.

The Book of Knowledge: Books 1 and 18. New York: The Grolier Society, Inc., 1942.

Buisseret, David. *Historic Illinois From The Air.* Chicago and London: The University of Chicago Press, 1990.

Calkins, Carroll C. *Mysteries of the Unexplained.* Pleasantville, NY: The Reader's Digest Association, Inc., 1982.

Childress, David H. *Lost Cities of North & Central America.* Stelle, IL: Adventures Unlimited Press, 1992.

Cirlot, J. E. *A Dictionary of Symbols.* New York: Dorset Press, 1971.

Clark, Ella E. "Quillayute." In *Indian Legends of the Pacific Northwest.* Berkeley: University of California Press, 1953.

Coleman, Loren. *Mysterious America.* New York: Paraview Pocket Books, 2007.

The Complete Books of Charles Fort. New York: Dover Publications, Inc., 1974.

Costa, David J. "Culture Hero and Trickster Stories." In Brian Swann, ed., *Algonquian Spirit.* Lincoln: University of Nebraska Press, 2005.

www.cryptozoology.com/articles/marlon.php

http://www.greatriverroad.com/Cities/Alton/PiasaBird.htm

Miller, Nancy K. "Tracking Pterosaurs." *Earth,* October, 1997.

MonsterQuest Episode 104: "Birdzilla."

http://www.mysteriousworld.com/Journal/1999/Autumn/Thunderbird/ default.asp

http://www.mysteriousworld.com/Journal/1999/Spring/Piasa01/default.asp

http://www.mysteriousworld.com/Journal/1999/Summer/Piasa02/default .asp

O'Conner, Mallory McCane. *Lost Cities of the Ancient Southeast.* Gainesville, FL: The University Press of Florida, 1995.

http://www.piasabirds.com/pisalegend.html

Schedler, Marcia. *Country Roads of Illinois.* Castine, ME: Country Roads Press, 1992.

Thompson, C. J. S. *Giants, Dwarfs, and Other Oddities.* New York: Citadel Press, Carol Publishing Group, 1968.

Welker, Glenn, ed. "The Origin of the Thunderbird." In *Indigenous Peoples' Literature,* retrieved at *http://www.indigenouspeople.org/natlit/natlit. htm.*

The WPA Guide To Illinois. New York: Pantheon Books, 1939, 1966, 1983.

Jake and Other Alligator Men

http://www.columbian.com/news/strange/outerlimits/alligator.cfm

http://www.marshsfreemuseum.com/pages/jake.html

http://www.roadsideamerica.com/story/2973

http://en.wikipedia.org/wiki/Colombian_folklore#Legendary_creatures

http://en.wikipedia.org/wiki/Jake_the_Alligator_Man

Beast of Bray Road and Other Contemporary Werewolves

http://www.americanmonsters.com/interviews/linda/interview.html

http://www.beastofbrayroad.com

Godfrey, Linda S., Richard Hendricks, Mark Moran, and Mark Sceurman, eds. *Weird Wisconsin: Your Travel Guide to Wisconsin's Local Legends and Best Kept Secrets.* New York: Sterling, 2005.

Guiley, Rosemary Ellen. "The Beast of Bray Road: A Modern Werewolf in America." Retrieved at *www.visionaryliving.com/2008/09/17/the-beast-of-bray-road-a-modern-werewolf-in-america.*

———. *The Encyclopedia of Vampires, Werewolves and Other Monsters.* Facts On File, 2004.

Hall, Jamie. *Half Human, Half Animal: Tales of Werewolves and Related Creatures.* Bloomington, IN: Authorhouse, 2003.

MonsterQuest Episode 114: "American Werewolf."

Moran, Mark, and Mark Sceurman. *Weird US: Your Travel Guide to*

America's Local Legends and Best Kept Secrets. New York: Barnes & Noble, 2004.

Opsasnick, Mark. "On the Trail of the Goatman." *Strange Magazine* 14, 1994.

http://www.prairieghosts.com/werewolves.html

Sankey, Scarlet, "Bray Road Beast." *Strange Magazine 10,* 1992.

http://www.unknown-creatures.com/bray road beast.html

http://www.visionaryliving.com/articles/beastofbray.php

http://en.wikipedia.org/wiki/Beast_of_Bray_Road

Goatman

http://www.geocities.com/Area51/Aurora/4746/feature2.html

http://www.goatmanhollow.com/the_legend/past_season_legends.html

Newton, Michael. "Goatman." In *Encyclopedia of Cryptozoology: A Global Guide.* Jefferson, NC: McFarland & Company, Inc., 2005.

"Residents Fear Goatman Lives: Dog Found Decapitated in Old Bowie." *Prince George's County News,* November 10, 1971.

http://www.unknown-creatures.com/maryland goatman.html

http://en.wikipedia.org/wiki/Goatman_%28Maryland%29

http://en.wikipedia.org/wiki/Satyr

Giant Catfish

http://www.angelfire.com/bc2/cryptodominion/fish.html

http://animals.nationalgeographic.com/animals/fish/mekong giant catfish .html

http://animal-world.com/encyclo/fresh/catfish/catfish.htm

http://www.aquaticcommunity.com/catfish/

Axelrod, Herbert R., C. Emmens, W. Burgess, and N. Pronek. "Exotic Tropical Fishes." Portland, OR: T.F.H. Publications, 1996.

http://www.dailylobo.com/index.php/article/2007/04/panicinducing_cat-fish_the_size_of_school_buses_make_waves_and_good_stories

http://www.geocities.com/cryptidwrangler/kycryptids2.html

http://www.ihoneida.com/features/081805oden.html

http://www.lochnessinvestigation.org/Catfish.html

http://www.meta-religion.com/Zoology/Extremes/giant_catfish2.htm

http://www.mississippirivermuseum.com/fame/jolliet.cfm

MonsterQuest Episode 108: "Gigantic Killer Fish."

http://www.news.nationalgeographic.com/news/2005/06/0629_050629_
    giantcatfish.html

http://www.planetcatfish.com/cotm/index.php

http://polishpoland.com/catfish_photographs.htm

http://www.popmatters.com/pm/column/46324/giant catfishonly a noodle
    away/

http://www.theatlantic.com/issues/97feb/catfish/catfish.htm

http://unusualdeath.blogspot.com/2006/08/on august 25th 1998 franc
    filipic 47.html

http://www.worldwildlife.org/who/media/press/2005/WWFPresitem802.
    html

http://en.wikipedia.org/wiki/Giant_Mekong_Catfish

http://en.wikipedia.org/wiki/Candiru

http://zipcodezoo.com/Animals/Z/Zungaro_zungaro/Default.asp

Moas (Giant Flightless Birds)

Anderson, A., "On Evidence for the Survival of Moa in European
    Fiordland." New Zealand Journal of Ecology 12, 1994.

http://www.hbtoday.co.nz/localnews/storydisplay.cfm?storyid=3760032

Heuvelmans, Bernard. On the Track of Unknown Animals. New York: Hill
    and Wang, 1959.

http://www.newanimal.org/moas.htm

"Reconstructing the Tempo and Mode of Evolution in an Extinct Clade
    of Birds with Ancient DNA: The Giant Moas of New Zealand."
    PNAS 102, 2005.

http://www.teara.govt.nz/TheBush/NativeBirdsAndBats/Moa/3/en

http://www.unexplained mysteries.com/viewnews.php?id=62976

Weidensaul, Scott. *The Ghost with Trembling Wings: Science, Wishful Thinking and the Search for Lost Species.* New York: North Point Press, 2002.

http://en.wikipedia.org/wiki/Moa

Worthy, Trevor H., and Richard N. Holdaway. *The Lost World of the Moa.* Bloomington, IN: Indiana University Press, 2002.

Shadow People

http://www.coasttocoast.com

http://ezinearticles.com/?The Shadow People&id=229341

http://www.ghostweb.com/shadow_people.html

http://www.monstropedia.org/index.php?title=Shadow_people

Palmer, Jeffrey. "The Shadow People." *EzineArticles*, June 26, 2006.

http://www.shadowpeople.org/

Black Dogs

http://www.authorsden.com/visit/viewArticle.asp?id=22494

http://www.bbc.co.uk/jersey/myisland/folklore/black_dog.shtml

http://www.cprs.co.uk/phantomhounds.html

http://www.indigogroup.co.uk/edge/bdogs.htm

http://www.mysteriousbritain.co.uk/england/n_yorkshire/trollers_gill.html

http://www.mysteriousbritain.co.uk/folklore/black_dogs.html

http://www.mysteriousbritain.co.uk/folklore/englishfolkapp.html

http://www.mysterymag.com/earthmysteries/index.php?page=article&subID=74&artID=269

http://norfolkcoast.co.uk/myths/ml_blackshuck.htm

Redfern, Nick. "Phantom Hounds of the Woods." *Fate*, August, 2007. Retrieved at *http://www.fatemag.com/issues/2000s/2007 08article3 .html*

http://www.squidoo.com/blackshuck

http://www.unknown-creatures.com/black dogs.html

Vance, Randolph. *Ozark Superstitions.* New York: Columbia University Press, 1947.

http://en.wikipedia.org/wiki/Barghest

http://en.wikipedia.org/wiki/Black_dog_(ghost)

http://en.wikipedia.org/wiki/Black_Shuck

http://www.wtvr.com/global/story.asp?s=6078107

Kappa (Japanese River Imps)

http://www.asahi-net.or.jp/~dp8h izn/gallery.html

http://www.brainyencyclopedia.com/

http://www.encyclopedia/k/ka/kappa___mythical_creature_.html

http://www.ffortune.net/spirit/zinzya/kami/kappa.htm

http://www.ishinotent.co.jp/Kappa/kappa 2.html

http://www.mangajin.com/mangajin/samplemj/ghosts/ghosts.htm

http://www.onmarkproductions.com/html/kappa.shtml

http://www.pinktentacle.com/2008/05/seven mysterious creatures of japan/

http://www.pref.iwate.jp/english/folklore/folklore.html

http://www.sagasaga/archives/sagasaga9809.htm

http://en.wikipedia.org/wiki/Kappa_(folklore)

http://www.yomiuri.co.jp/nanjo/nanjo13.htm

Encantados (Dolphin-Men of Brazil)

http://www.bbc.co.uk/nature/wildfacts/factfiles/62.shtml

Hall, Jamie. "Enchanted Dolphins." In *Half Human, Half Animal: Tales of Werewolves and Related Creatures.* Bloomington, IN: 1st Books, 2003.

http://monsterguide.blogspot.com/2006/02/encantado dolphin man of amazon river.html

Rice, Dale W. "Marine Mammals of The World: Systematics And Distribution." *Society of Marine Mammalogy Special Publication 4,* 1998.

http://en.wikipedia.org/wiki/Boto

http://en.wikipedia.org/wiki/Encantado

Clurichauns

http://www.bellaterreno.com/art/irish/fairy/irishclurichaun.aspx

Briggs, Katharine. "Leprechauns." In *An Encyclopedia of Fairies, Hobgoblins, Brownies, Bogies, and Other Supernatural Creatures.* New York: Pantheon, 1976.

http://www.geocities.com/Athens/Forum/4611/fairyC.html

http://www.monstropedia.org/index.php?title=Clurichaun

http://www.nationmaster.com/encyclopedia/Clurichaun

http://www.the-atlantic-paranormal-society.com/articles/naturespirit/compendium.html

http://en.wikipedia.org/wiki/Clurichaun

http://en.wikipedia.org/wiki/Fairy_ring

Yeats, W. B. *Fairy and Folk Tales of the Irish Peasantry,* In *A Treasury of Irish Myth, Legend, and Folklore.* New York: Gramercy, 1988.

Giant Spiders

http://www.ag.arizona.edu/yavapai/anr/hort/byg/archive/solpugid.html

http://www.arachnology.be/pages/Solifugae.html

http://www.badspiderbites.com/camel spider/

http://www.cryptomundo.com/cryptozoo news/giantspiders/

http://news.nationalgeographic.com/news/2004/06/0629_040629_camelspider.html

http://www.tinkertakeoff.com/article.htm?intRecID=1806

Boston Lemur

http://www.cryptozoology.com/sightings/sightings_show.php?id=588

http://www.newyorker.com/talk/2008/08/04/080804ta_talk_mcgrath

Dragons

http://www.cryptomundo.com

http://www.megafauna.com/chapter5.htm

http://www.museum.vic.gov.au/prehistoric/mammals/megalania_prisca
.html

http://www.mysteriousaustralia.com/amazing_creatures_lizards.html

http://www.parks.sa.gov.au/naracoorte/wonambi/animals/extinct/005780

http://www.unmuseum.org/bigliz.htm

http://en.wikipedia.org/wiki/Megalania

Spring Heeled Jack

http://www.bbc.co.uk/legacies/myths_legends/england/black_country/
article_2.shtml

Coleman, Jerry. *Strange Highways: A Guidebook to American Mysteries &*
the Unexplained. Alton, IL: Whitechapel Productions Press, 2003.

http://www.theparanormalreport.com/Spring Heeled Jack and the Devil in
Devon Revisited.html

http://en.wikipedia.org/wiki/Spring_Heeled_Jack

Beast of Bladenboro, North Carolina

http://www.cryptomundo.com/cryptozoo news/beastfest 08/

Jefferson, Paul. "Bolivia Predator Remains a Mystery." *Starnewsonline.com*,
October 26, 2007.

———. "Pit Bull is Mauled by Beast of Bolivia." *Starnewsonline.com*,
December 12, 2007.

MonsterQuest Episode 202: "Vampire Beasts."

http://supernaturalparanormal.com/?p=546

Ware, Jim. "A Scary Time for Trick Or Treating in Brunswick."
Starnewsonline.com, October 31, 2007.

Ohio Grassman

http://www.bigfootencounters.com/creatures/grassman.htm

http://www.history.com/shows.do?action=detail&episodeId=360320

MonsterQuest Episode 204: "Ohio Grassman."

http://www.today.com/external.php?url=http://feeds.feedburner.com/~r/ PhantomsAndMonstersAPersonalJourney/~3/310225856/ohio grass man to be featured on.html&reffurl=http://www.today.com/view/ohio grassman to be featured on monsterquest/id 2157249/

Lizard Man of Scape Ore, South Carolina

http://www.cnn.com/video/#/video/offbeat/2008/03/01/sc.lizard.man. update.wis

Horswell, Cindy. "Lizard Man Leaves Mark." *The Houston Chronicle*, July 30, 1989.

———. "'Lizard Man' Legend Still Alive, Kicking." *The Houston Chronicle*, July, 1989.

———. "On a Scale of One to 10, It Rates a Downright Scary 11." *The Houston Chronicle*, July 31, 1988.

"'Lizard Man' Claims a Casualty." *The Washington Post*, August 4, 1988.

Milligan, Stephen. "Sightings of a Monster Lizard from the Swamp have Struck Terror into a small Community in South Carolina." *The Sunday Times*, August 7, 1988.

MonsterQuest Episode 202: "Vampire Beasts."

http://www.skepticworld.com/cryptozoology/reptilians.asp

http://theshadowlands.net/creature2.htm#lizard

"To Keep a Monstrous Legend Alive, Man Admits Lying About Lizard Man." *The Houston Chronicle*, August 13, 1988.

http://www.wistv.com/Global/story.asp?S=7978464

"Youth Who Saw 'Lizard Man' Gets an Agent." *San Francisco Chronicle*, August 8, 1988.

Lobo (Wolf) Girl of Texas

Botkin, Benjamin Albert. *The American People*. Louis Filler, NY: Pilot Press, Ltd., 1946.

http://www.feralchildren.com/en/showchild.php?ch=felipe

Holstun Lopez, Barry. *Of Wolves and Men*. New York: Charles Scribner's Sons, 1978.

http://www.mysteriouspeople.com/Wolf_Girl.htm

Giant Rats of New York City

http://www.animalinfo.org/species/rodent/hypoanti.htm

http://floridakeystreasures.com/creatures/pouchrat/

"For Sniffing Out Land Mines, a Platoon of Twitching Noses." May 18, 2004. Retrieved at *http://www.acs.appstate.edu/~kms/classes/psy3202/RatsLandMines.htm*

MonsterQuest Episode 206: "Super Rats."

http://news.bbc.co.uk/2/hi/science/nature/7149569.stm

Novak, R. M., and J. L. Paradiso. *Walker's Mammals of the World, Vol II*. Baltimore: Johns Hopkins University Press, 1991.

http://www.nytimes.com/2008/01/17/science/17rat_web.html?_r=1&oref=slogin

Perry, N. D., et al. "New Invasive Species in Southern Florida: Gambian Rat (Cricetomys Gambianus)." *Journal of Mammalogy*, 2006, 262–264.

Peterson, A. T., et al. "Native Range Ecology and Invasive Potential of Cricetomys in North America." *Journal of Mammalogy*, 2006, 427–432.

http://www.timesonline.co.uk/tol/news/uk/science/article3193462.ece

http://en.wikipedia.org/wiki/Coypu

http://en.wikipedia.org/wiki/Cricetomys_emini

http://en.wikipedia.org/wiki/Gambian_pouched_rat

http://en.wikipedia.org/wiki/Giant_pouched_rat

Dover Demon

http://www.book of thoth.com/article1208.html

Clark, Jerome. *Unexplained!* Detroit: Visible Ink Press, 1999.

http://members.aol.com/soccorro64/doverdemon.htm

http://www.newanimal.org/dover demon.htm

Nyman, Joseph. "Investigation Into the Reports of an Unknown Creature Seen In Dover, Massachusetts April 21–22, 1977." January 28, 1978.

http://strangene.com/monsters/dover.htm

http://en.wikipedia.org/wiki/Dover_Demon

http://en.wikipedia.org/wiki/Mannegishi

Mothman of West Virginia

Coleman, Jerry D. *Strange Highways: A Guidebook to American Mysteries & The Unexplained.* Alton, IL: Whitechapel Productions Press, 2003.

Coleman, L. *Mothman and Other Curious Encounters.* 2007.

Colvin, Andrew. *The Mothman's Photographer: The Work of an Artist Touched by the Prophecies of the Infamous Mothman.* 2007.

———. *The Mothman's Photographer II: Meetings With Remarkable Witnesses Touched by Paranormal Phenomena, UFOs, and the Prophecies of West Virginia's Infamous Mothman.* 2007.

Keel, John A. *The Mothman Prophecies.* Saturday Review Press, 1975.

———. *The Mothman Prophecies.* 2007.

http://www.prairieghosts.com/moth.html

Sergent, Jr., Donnie. *Mothman: The Facts Behind the Legend.* 2001.

Taylor, Troy. "Unexplained America—Mothman: The Enigma of Point Pleasant." Dark Haven Entertainment.

http://www.westva.net/mothman/

http://en.wikipedia.org/wiki/Mothman

New Jersey Devil

http://www.aclink.org/history/mainpages/jerseydevil.htm

http://www.aclink.org/history/mainpages/jerseydevil2.asp

Cohen, David Steven. *The Folklore and Folklife of New Jersey.* New Brunswick: Rutgers University Press, 1983.

Coleman, Jerry D. *Strange Highways.* Alton, IL: Whitechapel Productions Press, 2003.

Hauck, Dennis William. *The National Directory of Haunted Places.* New York: Penguin, 1996.

Jagendorf, M. A. *Upstate, Downstate: Folk Stories of the Middle Atlantic States.* New York: Vanguard Press, 1949.

http://jerseyhistory.org/legend_jerseydevil.html

McCloy, James F., and Ray Miller, Jr. *The Jersey Devil.* Walingford, PA: Middle Atlantic Press, 1976.

http://www.prairieghosts.com/jerseydevil.html

http://www.strangemag.com/jerseydevil1.html

http://theshadowlands.net/jd.htm

http://en.wikipedia.org/wiki/Jersey_Devil

Chupacabra

http://classes.colgate.edu/gned339/Chupa/

Coleman, Jerry D. *More Strange Highways.* Alton, IL: Whitechapel Productions Press, 2003.

http://www.crystalinks.com/chupacabras.html

http://www.ebe.allwebco.com/news/

http://www.geocities.com/Area51/

http://www.geocities.com/area51/Shadowlands/6583/cattle031.html

http://www.kvue.com/news/top/stories/073107kvuechupacabrafind cb.cc11e691.html

MonsterQuest Episode 208: "Chupacabra."

Neer, Catherine. *How Chupacabras Work.* Retrieved at *http://science. howstuffworks.com/chupacabra.htm*

http://www.prismnet.com/~patrik/chupa3.htm

http://en.wikipedia.org/wiki/Chupacabra

http://zetatalk.com/theword/

A Tale of MoMo (Missouri Monster)

Gilbert, Joan. *Missouri Ghosts*. Hallsville, MO: MOGho Books, 2001.

http://www.newanimal.org/momo.htm

http://www.stateofhorror.com/momo.html

http://en.wikipedia.org/wiki/Momo_the_Monster

About the Author

Photograph © Brittany Budd

Deena Budd lives in Hannibal, MO, with her dog, Goku. She writes weekly articles for www.bella-online. com, where she serves as Paranormal Editor. She has also worked in radio for many years and hosts a Monday morning show on KHBL 96.9 out of Hannibal. When she's not writing or hosting, she can be found reading, hiking, caving, driving her '66 Buick Wildcat, and exploring any thing and any place that looks intriguing